ROYAL YACHTS
Under Sail

Royal Yachts
Under Sail

Brian Lavery

Seaforth
PUBLISHING

Half title: Vincenzo Coronelli's drawing of a royal yacht, from his book *Ships and Vessels*, Venice 1697. See page 129.

Title page: Detail of a painting by Peter Monamy depicting the arrival of George I in the lower Thames. See page 59. (Richard Green Gallery)

Copyright © Brian Lavery 2022

First published in Great Britain in 2022 by
Seaforth Publishing,
A division of Pen & Sword Books Ltd,
47 Church Street,
Barnsley S70 2AS
www.seaforthpublishing.com

British Library Cataloguing in Publication Data
A catalogue record for this book is available from the British Library

ISBN 978 1 3990 9291 3 (HARDBACK)
ISBN 978 1 3990 9292 0 (EPUB)
ISBN 978 1 3990 9293 7 (KINDLE)

All rights reserved. No part of this publication may be reproduced or transmitted in any form or by any means, electronic or mechanical, including photocopying, recording, or any information storage and retrieval system, without prior permission in writing of both the copyright owner and the above publisher.

The right of Brian Lavery to be identified as the author of this work has been asserted by him in accordance with the Copyright, Designs and Patents Act 1988.

Pen & Sword Books Limited incorporates the imprints of Atlas, Archaeology, Aviation, Discovery, Family History, Fiction, History, Maritime, Military, Military Classics, Politics, Select, Transport, True Crime, Air World, Frontline Publishing, Leo Cooper, Remember When, Seaforth Publishing, The Praetorian Press, Wharncliffe Local History, Wharncliffe Transport, Wharncliffe True Crime and White Owl

Typeset and designed by Mousemat Design Ltd

Printed and bound in India by Replika Press Pvt Ltd

~ Contents ~

	Acknowledgements	6
I	**Sailor Kings**	7
1	The Origin of the Yacht	7
2	The First English Yachts	13
3	The Uses of the Yachts	19
4	The Later Yachts	27
5	The Loss of the *Gloucester*	34
6	The Yachts at War	38
II	**To the Wars**	44
1	The Changing Role	44
2	New Yachts	50
3	Marlborough at Sea	56
III	**The Hanover Connection**	58
1	The Coming of King George	58
2	The Yachts at Deptford and Greenwich	63
3	Rebuild, Repair, Renovation and Replacement	69
4	Other Duties	72
5	The Last King in Battle	77
IV	**Marriage, Business and Leisure**	83
1	A New Queen	83
2	Sandwich's Visitations	87
3	An Unhappy Marriage	92
4	The Yachts at Weymouth	94
5	New Yachts for a New Century	99
6	The End of the Wars	103
V	**Uniting the Kingdom**	107
1	George IV in Ireland	107
2	To Scotland	111
3	The Lord Admiral's Excursion	115
VI	**Building and Sailing**	118
1	Captains and Crews	118
2	Design and Build	124
3	Accommodation	128
4	Art and Decoration	133
5	Sailing the Yachts	139
VII	**Last Days Under Sail**	145
1	Victoria in Scotland	145
2	Second Wind	149
	Bibliography	154
	Notes	155
	Index	158

Introduction

My first, very indirect, contact with royal yachts came in the spring of 1953 when the young Queen Elizabeth II was driven past our school on the way to launch the *Britannia* in Clydebank. More than forty years later I attended one of the last receptions on board her in Portsmouth. The decommissioning of her shortly afterwards remains controversial.

There have been several very good books on royal yachts already, as listed in the bibliography. This one, I hope, probes deeper than before into the design and construction of these vessels, aided by the detailed cataloguing of the Navy Board letters in the Adm 106 in the National Archives, which opens up many possibilities for research, and by a number of recent articles in *The Mariner's Mirror*. I first became interested in the subject after researching a Peter Monamy painting on behalf of the Richard Green Gallery in London. It showed, as it turned out, the arrival of George I in England in 1714, one of several examples of British and French regime change facilitated by the yachts. This led to the recognition that very little had been written about the actual uses of the yachts. Broadly, they served mainly as pleasure craft, diplomatic vehicles, transports of kings and military leaders in peace and war, and occasionally as war vessels in their own right. Steam yachts opened up new possibilities, allowing Queen Victoria to visit the more remote parts of the United Kingdom, and Elizabeth II to have a kind of floating palace as a base during overseas visits.

The original yacht, of Dutch origin and British by adoption, developed in at least two different directions, which is reflected in the double meaning of the word today. It can be a sailing vessel, usually decked and with simple accommodation, used for cruising or racing by people of quite moderate means; or it can be a vastly expensive and luxurious vessel owned by a billionaire, and probably used more as floating accommodation than as a means of getting from A to B.

Their potential uses prompts the recurring question: should a new British royal yacht be built? Clearly the world is very different, not just from 1660 when Charles II first encountered yachts on his ecstatic passage back to England, but from the 1950s when the *Britannia* was designed and built. Britain no longer has an empire, and the sea is no longer an important means for people to travel long distances, except for leisure or as desperate would-be immigrants. But more important than that, it is difficult to imagine any government paying for a yacht which could match the opulence of those belonging to the plutocrats; a new *Britannia* would not be guaranteed to enhance British prestige.

Instead, I recommend readers to savour the yachts of history, as represented in models, an exquisite print by Vincenzo Coronelli, and the paintings and drawing of such masters as the Van de Veldes, Sir Godfrey Kneller, Peter Monamy, Dominic and John Thomas Serres, J C Schetky, and Nicolas Pocock; and to relish their fascinating history as revealed by documents and manuscripts of the past.

Brian Lavery, May 2022

Part I

Sailor Kings

1

The Origin of the Yacht

The idea of a specialised ship for royal transport did not originate until well into the modern age. Medieval English monarchs usually travelled short distances by sea to visit their realms, especially their holdings in France, and initially this was organised on a feudal basis. In 1240 the lord of the manor of Padworth in Berkshire, for example, held the manor in exchange for finding a man to hold a rope in the Queen's ship when she crossed from England to Normandy.[1]

Otherwise kings mostly travelled for warlike purposes. For England the medieval era began with William the Conqueror's voyage across the English Channel in 1066. He headed a fleet of around seven hundred ships, apparently in a standard open Viking-type longship like many others. According to the Bayeux tapestry it was distinguished by a rectangular structure at the head of the mast, which may have been a beacon, as fire was one of the signals available to him. It was faster than most, however, and at one stage he almost jeopardised the operation by going too far ahead. He attempted to reassure his men by having 'a plentiful and cheerful meal, as if he was in his own hall, served with spiced wine', suggesting that he had at least some facilities on board.[2]

The longest voyage by a medieval English king was Richard I's crusade to the Holy Land in 1190. He joined his fleet at Marseilles, leaving his ally Philip Augustus of France, who suffered from seasickness, to travel by land through Italy. Richard settled in a galley which was one of a group sailing a fleet of more than a hundred ships, 'with a small household, and the chief men of the army, with his attendants'. We know no more about his accommodation on board, but as in most royal voyages there was as much show and pageantry as possible: 'So great was the splendour of the approaching armament, such the clashing and brilliancy of their arms, so noble the sound of the trumpets and clarions'.[3]

Henry VIII sets out from Dover for the Field of the Cloth of Gold, 1620. His bulky form can just be seen on the ship with golden sails in the distance to the right. (Royal Collection Trust / © Her Majesty Queen Elizabeth II 2022, 405793)

In 1415 Henry V mounted an even larger expedition of at least eight hundred ships to invade France. He was to travel in the recently rebuilt *Trinity Royal*, one of his largest ships at about 540 tons. She had at least three cabins, which were quite rare at the time, but apparently that was not enough. She was known as the King's Chamber and another ship (perhaps the *Holy Ghost*) as the King's Hall, implying that he used one ship for sleeping and the other for the business of organising his fleet before it sailed from the Solent.[4] Again, there was as much show as possible. The flagship had a huge purple sail with a cover bearing the King's badge, and the emblem of the Trinity on a standard from the mainmast. There were long pendants or streamers bearing the arms of St Edward and St George, the top-castle had a large gilt copper crown, the prow was decorated with a golden leopard, and even the functional capstan had a fleur-de-lis emblem emphasising Henry's claim to the throne of France.[5] Henry VIII's trip to the famous Field of the Cloth of Gold in 1520 was ostensibly for more peaceful purposes. The maritime part was only the short trip from Dover to Calais, and any grandeur was greatly overshadowed by the elaborate and expensive decorations of the 'field' itself. Henry was depicted on the deck of one of his ships as it left Dover at the centre of a fleet bearing flags and shields – though not much more than they did on more ordinary occasions.

Luxury and splendour

Medieval kings knew little of luxury as it would be interpreted by their successors. They usually had a peripatetic existence, travelling between one fortified castle and another. They were expected to be war leaders and might spend a large proportions of their lives on campaign, living in tents. Even in castles they had little privacy, eating communally in the great hall. Furniture was usually moveable, including beds, wardrobes, chests and tapestries for decoration. It was not too difficult to transfer such facilities to the cabin of a ship if necessary. Tudor monarchs did not progress much beyond this, living in palaces such as St James's, Hampton Court, and Placentia at Greenwich – all of which consisted of castellated buildings round courtyards. The early Stuart kings, James I and Charles I, began to change this. They employed Inigo Jones to replace Placentia with the Queen's House at Greenwich, the first Palladian building in England, and build the Banqueting Hall among the rambling Whitehall Palace. By the middle of the seventeenth century new standards of grace and comfort were being set.

At the same time there was an explosion in ship decoration, largely because insecure monarchs needed ways to express their power and grandeur. The decorations of Henry V's *Trinity Royal* were largely ephemeral and would probably be put into store as soon as the royal presence ended. Tudor ships had relatively moderate decoration, perhaps a small figurehead and a few carved rails and balusters. They largely relied on paint and banners to express any glory. It was very different for Charles I's great ship, the 100-gun *Sovereign of the Seas* of 1637. It was festooned with carvings on bow, stern and sides and described by the poet Thomas Carew as 'the eighth wonder of the world'.

Such grandeur was reflected in France with King Louis XIII's flagship *La Couronne*, but it was very different in the Netherlands (usually known as the United Provinces at the time), which was almost a republic, with only the stadtholder as a hereditary office, with limited and uncertain power. Warships were supplied by the East India companies and the individual provinces, which did indeed decorate them, but not to the extent of the kings to the west and south. Instead, much of the effort was at a lower level, in the small craft or 'yachts' built by local corporations or private individuals to show off

The young Charles II drawn by Wenceslaus Hollar, c1650.
(Royal Collection Trust / © Her Majesty Queen Elizabeth II 2022, 803492)

their wealth. They were helped by the great flourishing of Dutch culture, especially painting, with the works of Rembrandt and Vermeer among many others. Exquisite decoration was to be a feature of Dutch yachts.

At the highest level were the States yachts, official vessels used by the members of the States General or the provincial parliaments. They were usually single-masted and fore-and-aft rigged. The spritsail with the boom running diagonally across the sail was being replaced by the gaff rig with the boom at the head of the sail. They had shallow draught for the numerous inland waterways of the country, and large leeboards to prevent them being driven sideways. Each had a large stateroom aft, wide because of the transom at the stern, there was another grand room midships, and a pantry and galley in the forecastle. The privately owned version was known as the burger yacht. The boier, or kop yacht, was a smaller vessel with rounded bow and stern in typical Dutch style. And a print of around 1640 showed an open vessel with two masts, described as a *speel yacht*, or pleasure yacht. It was presumably the property of a merchant, for the print has the apposite inscription: 'Earnings which are often increased by sailing in trade / Are consumed again in sailing for pleasure.'[6]

The Prince of Wales and the sea

Charles Stuart, Prince of Wales, had his first sea experience in 1646. His father, King Charles I, had been fighting a civil war against Parliament for four years and his forces were being driven back towards the extremities of England. After some time in the west of England the prince, then fifteen years old, was sent to the Scilly Isles. He was fascinated during the voyage in the frigate *Phoenix* and insisted on taking the helm himself. A few weeks in the islands allowed some chance to learn techniques in local craft, then he was sent to the Channel Island of Jersey, which was regarded as safer. Again he took the helm of the *Proud Black Eagle*, and once there he had further opportunity for local sailing among a seafaring people. A boat was built for him in the nearby French port of St Malo. He also had his first romance, setting another theme for his life that would also be reflected in his royal yachts.

Charles went on to France while his father was defeated at the Battle of Naseby and held prisoner on the Isle of Wight. In 1648 the prince had the chance to lead a rebellion by the Scots and began to make his way there by sea. He encountered a section of the Parliamentary fleet which had revolted against naval discipline, and he had the exhilarating experience of winning their loyalty for a short time. A naval battle in the Thames Estuary seemed imminent at one point, and Charles bravely refused to go below during it. 'Many officers and sailors of the admiral [ie flagship] beseeching the Prince to go into the hold to save himself, but his Highness told them that he could not preserve his honour which was dearer to him than his life'.[7] But the ill-disciplined sailors declined to fight and the ships melted away. In the meantime, the Scottish revolt had been crushed. Charles took refuge in Holland, where yet again he had a chance to see the customs of a seafaring people. It was probably then that he first encountered 'great barges, commonly called yachts … a kind of little frigates, whereof persons of condition make use upon the rivers, in passing from one province into another, for necessity, or for divertissement.'[8]

Escape by sea

Another opportunity for Prince Charles arose in Scotland in 1651 and he sailed there. The Presbyterian rulers demanded almost impossible conditions of him, but nevertheless he led their army south, only to be defeated at Worcester. His escape involved hiding up the famous oak tree (a story he never tired of telling). In October he arrived at the small port of Shoreham on the south coast of England. His supporters found the *Surprise*, a collier of under 60 tons, with a crew of four men and a boy, commanded by Nicholas Tattersall (variously spelled). At four in the morning of the 15th Charles and Lord Wilmot climbed a ladder to board the coaster which was beached at low tide. Charles lay down on the bed and Tattersall came below to tell him that he knew who he was, for he had been stopped by Charles's ship during his command of the mutinous fleet in 1648. He assured the King that he 'would venture life, and all that he had in the world, to set me down safe in France.'[9]

They sailed on high tide at seven o'clock that morning and headed towards the Isle of Wight, maintaining the fiction that they were delivering coal to Poole. The crew was told that the two passengers were merchants avoiding arrest for debt and trying to collect money owing to them in Rouen. At five in the afternoon, with a fair northerly wind behind them, they duly turned south and were off the French coast near Fécamp by daybreak. The wind shifted to the southwest and the tide was falling so they anchored two miles offshore. Charles spotted a well-handled ship, which he thought was a privateer from Ostend in the Spanish Netherlands prosecuting a war against France. Fearing capture and imprisonment or return to England, they persuaded Tattersall to have them taken to the land in the *Surprise*'s cock-boat. Once there, Charles was carried ashore on the shoulders of the mate Richard Carver, a Quaker. The 'privateer' turned out to be a harmless French hoy, and her captain was probably never aware of Charles's unwitting compliment to his seamanship.

Charles entering Rotterdam to an ecstatic reception, painted by Lieve Pieterz. The King is the figure with dark hair and dark face standing just aft of the sail on the yacht. (Rijksmuseum)

Opposite above: The *Surprise* after she was taken into the navy at the Restoration of 1660, and renamed *Royal Escape*. (Royal Collection Trust / © Her Majesty Queen Elizabeth II 2022, 405211)

Opposite below: Dutch vessels *c*1655 as drawn by Willem Van de Velde the Younger, with a typical yacht to the left. (Royal Collection Trust / © Her Majesty Queen Elizabeth II 2022, 405328)

~ Charles's fortunes change ~

After that, Charles endured years of depressing and penurious exile, being driven from Paris to Cologne, then to Brussels and Bruges. There was a ray of hope in 1658 when his arch-enemy Oliver Cromwell, Lord Protector of England, died suddenly. His son Richard was not up to the job and the country went through a series of governments until it was finally decided to restore the monarchy in the person of Charles. He went to Breda, where he issued a declaration promising a certain amount of religious toleration, respect for property acquired legally, and a pardon for his enemies, except those who had signed his father's death warrant.

The Dutch were keen to gain favour with the new king, and escorted him to Moerdijk on the Hollands Diep, one of the mouths of the Rhine. They had assembled a fleet of thirteen yachts plus pinnaces and other vessels 'almost innumerable'. A bridge had been constructed at the end of the dyke for the King to board, but immediately he declined the yacht offered for his accommodation. The boat of his sister, the Princess Dowager of Orange, was less commodious than that belonging to the Admiralty of Rotterdam, and Charles insisted in using the latter, which he had used before – it was 'without doubt, the greatest of all that little fleet'.

Charles's brother, James, Duke of York, agreed to 'perform the functions of admiral', that is, allocate places among the other yachts, the Dutch officials having declined to enter such a minefield. James took the Princess Dowager's yacht for himself and gave his other brother, the Duke of Gloucester, that of the Estates of Holland, while their sister had that of the Council of State. The yacht belonging to Mr Beverweert, the representative of Amsterdam, was allocated to the deputies of the Estates of Holland, while Charles's principal adviser, Sir Edward Hyde, now Lord Chancellor of England, went in a pinnace called *Maid of Zealand*. Room was found for various other lords and officials from both governments, including the marquesses of Ormond and Worcester.

Much comfort was on offer:

> Every yacht had its steward, and all other officers necessary for the kitchen and buttery, and they which had not the commodity to have the kitchen aboard themselves were accompanied with other barks, where chimneys were made for the kitchen and ovens or pastry and provision of so prodigious quantity of all sort of meats, of foul of sweet meats, of wine, that all the tables were perfectly served therewith …

THE ORIGINS OF THE YACHT

On the beach at Scheveningen, Charles says goodbye to Elizabeth of Bohemia and embarks in a naval barge to the English flagship to sail to Dover. (© National Maritime Museum, Greenwich, London, PAH1725)

It was a revelation to a Prince who had spent his youth in flight or poverty. Moreover, '[t]he English stewards … were astonished thereat, and considered that they could not comprehend, how they could make ready in boats and agitation, twenty or five and twenty great dishes for every table.'[10]

The cavalcade sailed and dinner was prepared for noon, but even in the narrow waterways there was some motion and the Princess of Orange was seasick. The King enquired 'if there was means to shelter them, somewhat, under some rising land', and was told that they would reach Dort (Dordrecht) in an hour and a half. Actually it took more than three hours, but the townspeople were ready with their welcome. 'The rampart, and quay, were bordered with citizens, which were put into arms, and with a battery of great guns, which made many volleys, as well as the muskets, whilst the fleet passed there.' They intended to stop for the night about a quarter of a league, or less than a mile below the town, but Charles's enjoyment of the adulation and his delight in the yacht conflicted with his desire to get home to take up his crown.

News arrived that Parliament had unconditionally accepted his possession of the crown, and that a fleet under Admiral Montagu was not far off the coast, ready to take him across. This disrupted the Dutch plans, but at five o'clock in the afternoon the anchors were weighed and the fleet passed Rotterdam, tacking several times in a wider channel and with a contrary wind. Again the citizens fired their salutes every time the tack brought the King's yacht closer to the town. At Delftshaven they entered the narrow River Schie and finally stopped at Overschie, some miles up. They sailed at dawn next morning to reach Delft, not at its most glorious after fire damage. The highest royalty and officials of both nations met on board the yacht, then they proceeded by land towards The Hague. After being feted there for several days, the royal party met an English fleet of gloriously fitted and newly decorated men of war. On 23 May the King and his suite boarded the former republican flagship *Naseby*, which was quickly renamed *Royal Charles*. They sailed to Dover to be greeted by an ecstatic populace.

2
The First English Yachts

The *Royal Escape*

In a sense the *Royal Escape* might claim to be the first English royal yacht. She was the *Surprise*, in which Charles had escaped from Shoreham. After the Restoration, Charles bought her from Nicholas Tattersall and she was renamed. She spent most of her time moored in the Thames off Whitehall Palace, whether as a warning of the King's precarious position, or as evidence of him being favoured by fate. Tattersall, despite his humble origins, was commissioned as a naval captain in July 1660, then awarded a pension of £100 for life in 1663; he died nine years later and was buried in Brighton. The *Royal Escape* put to sea in 1672–74 during the Third Dutch War and ended up at Deptford Dockyard, where she survived in various forms until 1750. There is no sign that the royal family ever sailed in her after 1651 and she was usually listed as a smack, only occasionally as a yacht. Drawings and paintings by Van de Velde[1] show her with a window and gunport on each side, which was probably not part of her original layout as a collier; but apart from that she gives us the best information we have on a small trading vessel of the mid seventeenth century – 30ft 6in long, 14ft 3in broad (much wider than a warship or yacht of that size), 7ft 9in deep in the hold and drawing 7ft of water. She was measured at 34 tons, which gives some indication of her carrying capacity.[2]

The *Mary*

Charles had been impressed with the Amsterdam yacht during his procession through the Dutch waterways in 1660: 'And indeed the King found his yacht so fit, and so well fashioned, that he said in discourse with the deputies, that he would cause one to be made of the same manner, as soon as he arrived in England, to serve him upon the Thames above the bridge.' The last passage is highly unlikely: it would be impossible to get a vessel of that size between the piers of old London Bridge. In any case:

> Mr de Vlooswick, Burgomaster of Amsterdam, and one of the deputies of the Province of Holland, taking occasion from thence to render a very considerable service to this country, said to the King, that lately they had made one in their town of the same bigness, at least as commodious in every way, which he took the liberty to offer to his Majesty, beseeching him to grace the magistrate of

The yacht *Mary* as drawn by Van de Velde the younger, showing the Dutch style with leeboards and high, decorated stern. (Media storehouse)

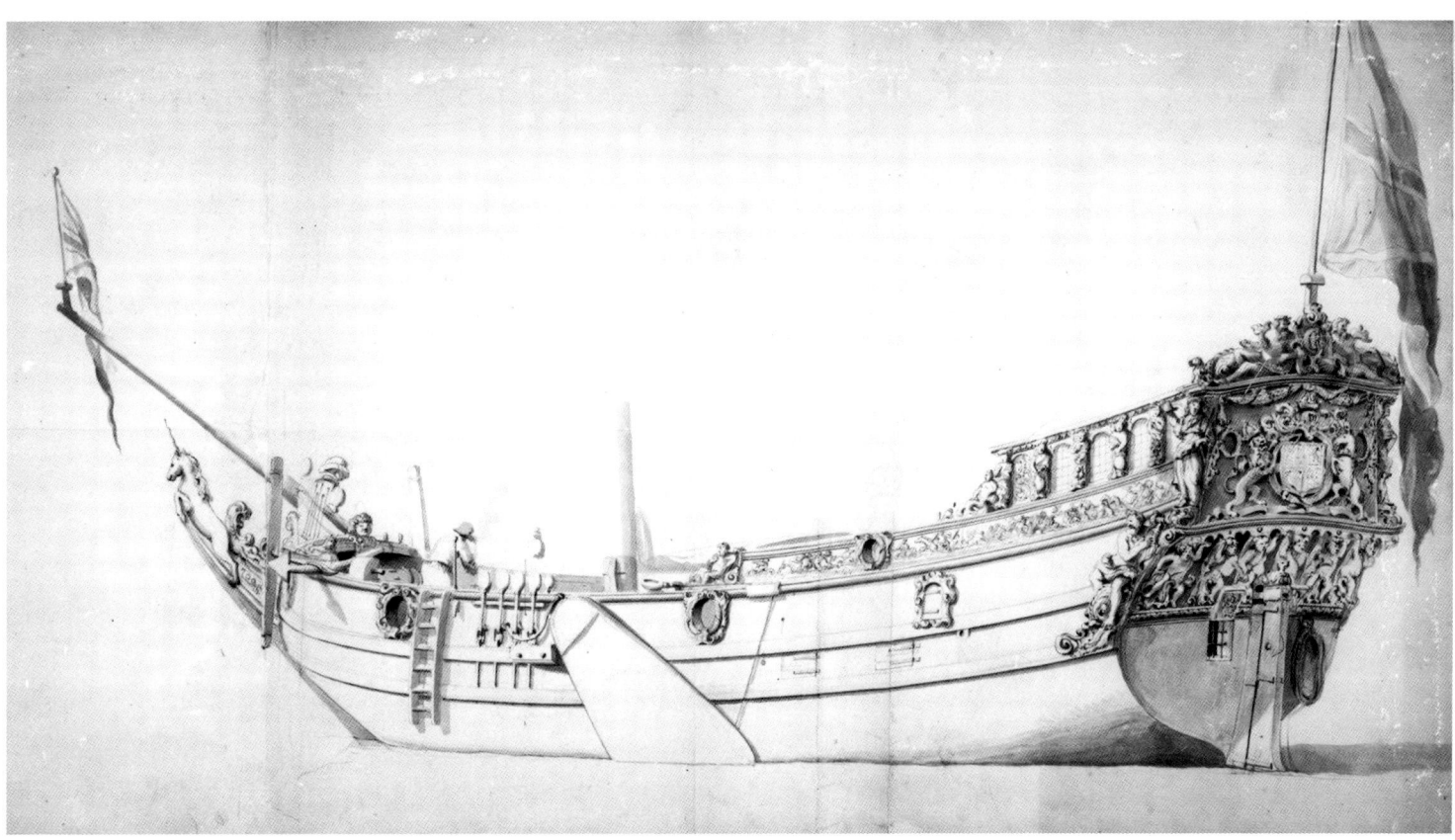

the town of Amsterdam to accept it. The King accepted it not absolutely, but declined not so strongly that upon the advertisement which Mr de Vlooswick gave to the magistrate of what passed on that occasion, he caused not the yacht to be bought which the College of Admiralty had gotten of the East India Company, and put it in a condition to serve for the diversion of this great Prince; and to give it the more lustre, the magistrate caused the outside to be richly gilt, whilst some of the best painters of the country wrought upon the fair pictures, wherewith thy have since adorned the inside.[3]

The burgomasters judged that yacht 'not a present worthy of his Majesty'. However, '[n]otwithstanding if his Majesty would be pleased to accept it, it would necessary to send someone, to the place, to order the contrivances, and accommodation, as for their part they would endeavour to give all the embellishments which might render it pleasing to his Majesty'. The King replied that he would indeed value such a yacht, but did not want to be obliged to the City of Amsterdam; but he would send the captain of Beverweert's yacht which had had carried him to Breda, with 'order to cause that to be finished'.[4]

The yacht was described by the 'Honourable Gentlemen Burgomasters' of Amsterdam.

> The yacht length across stem and stern 72 [Dutch] feet, width inside the planking 18 feet and 10 inches. Depth below the main deck 8 feet. Planked with three-wrist-thick boards, and the ceiling, with carvel planks. Furthermore the keel, stern post, stem, frame timbers, with beams and knees as it should be done. Aft a main cabin with 4 box beds, and couches, and aft [of that] a cabin for the captain. By the mast a room for the servant, and a galley. Forward a cabin for the sailors, and aft a pavilion. Furthermore rudder, lee boards, main deck, bollard timbers and wales; berths for the sailors; oakum, pitch and tar for as much as runs below the waterline.[5]

A drawing by Willem van de Velde the Younger, presumably made before she sailed for England, shows her with a typical Dutch square transom stern below, and prominent leeboards to be dropped and prevent her being pushed sideways by the wind, as in a Thames barge of later years. She has a narrow, angular head with a carving, presumably of a unicorn, with two wreathed gunports on each side and a porthole, but no quarter gallery. Her stern is high, with six prominent figures on each side, and presumably windows between them. The flat of her stern bears the royal coat of arms. She has a single mast supported by shrouds attached to very regular deadeyes and a steep bowsprit.[6]

Samuel Pepys went on board the *Mary* in November 1660 and she was 'one of the finest things I ever saw, for neatness and room in so small a vessel'[7] – a defining skill of the yacht builder ever since. She was built in Dutch fashion with a shallow draught and leeboards, but soon Charles began to build his own in English style with a deeper hull.

Pepys records her draught of water as 10ft, which is clearly absurd unless it was measured with the leeboards down. It was several feet more than any of the English-built yachts, and she would certainly not have been chosen for the shallow waters of the Dee Estuary with such a draught. The Burgomasters' report mentions a depth of 8ft below the main deck (0.93 of an English foot or about 7ft 5in in total). Drawings suggest that her main deck was 3ft or 4ft above the water, which implies a draught of around 3ft or 4ft, which is consistent with normal Dutch practice.

The year 1660 was not the first time the English had heard the word 'yacht', however spelled. The accounts of Hurstmonceaux Castle in Sussex for 1643/4 contain an item for making a dock at Pevensey 'for my lord's yought', and five shillings for a carpenter doing six days' work 'about the yaught'.[8] In the spring of 1656 the commander of a 'yatcht' complained about being stopped and searched by an English frigate, even 'searching also the earth tubs, where the lemon and orange trees were in, besides all the cabin's corners, and desks in the whole yacht'.[9] But the arrival of the *Mary* raised the word to a much greater prominence. The shipwright Sir Anthony Deane told Samuel Pepys that 'no Englishman could have built such a vessel for the sea and accommodation both, and so small, as a yacht before the *Mary* Yacht was brought from Holland.'[10]

Once the English yachts were in commission, the *Mary* found a new role in the service of the Lord Lieutenant of Ireland, transporting him and his associates to England and Wales. Her shallow draught was an advantage here, especially in the Dee estuary leading up to the port of Chester. Nevertheless, she was wrecked in 1673. Investigation of the wreck has not revealed much about her hull, but many artefacts have been recovered.

The *Katherine*

On 3 November 1660, five months after Charles's restoration, the Navy Board received an estimate to build a new yacht in the Royal Dockyard at Deptford, just a few miles downriver from London. It was to be of 80 tons and cost £1935. The *Katherine* of 1661 was to be built by Peter Pett, one of the ubiquitous family of shipwrights, and not to be confused with his cousin of the same name who became the scapegoat for the

The yacht *Katherine* as seen from the stern in 1672, with the *Cleveland* broadside on, by or after Van de Velde the Younger. (© National Maritime Museum, Greenwich, London, BHC0299)

Dutch raid on the Medway in 1667. Pett lodged near the Globe Tavern in Deptford while she was building, and was joined by Pepys for dinner in January 1661. He commented that the yacht would be 'a pretty thing, and much beyond the Dutchman's'.[11] The yacht was approaching completion by 3 June, when the King ordered a new suit of sails and Pett asked the Navy Board to bargain for the cloth required, and put them in hand. Four tons of musket shot were needed for ballast and it was delivered by August. The gilding was undertaken by Sir Robert Howard at a cost of £450, which must have made a substantial hole in the £1935 originally estimated. Sir Robert was Sergeant Painter to the King, a younger son of the Earl of Berkshire, a gallant Cavalier during the Civil War, and a politician as well as a painter. John Davis provided two pieces

The yacht *Katherine* of 1661 showing a more English style, also by Van de Velde. (© National Maritime Museum, Greenwich, London, PAH3912)

of red damask, 'being part of the furniture belonging to the King's yacht *Katherine*.' She was launched in 1661 and measured at 94 tons, 49ft long, 19ft broad and 7ft deep in the hold, slightly smaller than the *Mary* except in the breadth, which suggests that she was designed for sailing with the wind on the beam. She carried ten brass 3pdr guns, mainly for answering salutes.

The shipwright Nicolas Witsen had a chance to look over the *Katherine* after she was captured by the Dutch in 1673, and he provides details of the changes from the Dutch type. Most fundamentally, she was designed for seagoing rather than inland waterways, being lower in the water and 'so shaped and built that it can keep the sea' – at that time the English had practically no canals, only winding and highly tidal rivers. She was designed for deeper waters so she had a greater draught, and did not need the leeboards. The galleries in the stern were lower, 'since they keep the sea more than we are wont to do with ours'. Her sides were straight by Dutch standards, 'not of such a curved shape as those that are built in this country, but more straight-sided, which nevertheless does not look bad, though many masters consider this the most important feature of a ship.' The stem was 'upright', which meant that it was straight though angled sharply forward, and it was formed of a single piece of timber. The bow was wide above, 'but below, it is sharp and of a good shape, very proper to cut the water and make good speed.' The lower stern was rounded in English style, but above that it spread out 'high and broad'. The deck was almost flat. She was built, according to Witsen, 'of Irish timber … the finest wood to be found in Europe.' Because of this she was not tarred like a Dutch yacht, which showed 'all the better its nature and ruddy colour which is agreeable to the sight.' In principle, the rigging was unchanged from the Dutch style. 'Sail, mast and gaff of the shape that is usual in this country', but intriguingly he reveals that the mast could be moved 'to and fro to seek sailing power'.

The *Katherine* was returned after the peace of 1674, but given to the Board of Ordnance.

The *Anne*

James's yacht, the *Anne* (sometimes known as the *York* after his dukedom), was built downriver in the dockyard at Woolwich by Christopher Pett, another cousin. According to Anthony Deane's figures she was a foot longer than the *Katherine* on the keel and 4in narrower. Her draught of water was 7ft, 2in less than the *Katherine* and the same as the *Mary*. All that suggests that she was designed for downwind sailing, and the *Katherine* would be faster with the wind on the beam, which might explain the results of the famous race of 1661.

For some reason her building was more troublesome. In the last days of 1660 Pett was complaining about the shortage of planks for building her, and wanted the purveyor to be quickened. On 12 March 1661 he wanted 6 tons of old shot from the Tower of London for ballast, as the duke wanted the yacht to be launched in six weeks. He also wanted lead from Chatham, but he had to go through the Board of Ordnance, which took its usual bureaucratic approach. At the end of the month there was a serious dispute with the carvers, Thomas Eaton and Richard Swain. They should be 'severely punished for contempt; if they continue to refuse to work; they know the great necessity there is for them, and that without them the vessel cannot be finished by the time prefixed.' In the meantime, the ship was being fitted by the 'plumberess', Mrs Browne, probably a widow carrying on her late husband's business, rather than a practical craftswoman. On 12 April the duke ordered the yacht to be launched the following Thursday, though the joiners and carvers work were not complete. Sails had already been ordered in February, similar to those of the *Katherine*. The master painter Mr Walker was to adorn the ship at the comparatively modest cost of £160.

The *Henrietta*

King Charles was pleased with the *Anne*. On 18 May 1663 he visited Woolwich Dockyard and Commissioner Christopher Pett produced an estimate for a new yacht like her. The cost was to be £1850 without guns, less than *Katherine* and *Anne* because her cabins would be lined with leather, rather than carved work. This kept the cost down to £686 for the hull, or £6.35 per ton compared with £7 for other yachts.[12] The King continued to take an interest, first in her keel. One provided by a Mr Blackberry turned out to be rotten and another was obtained from William Graves. Pett listed the timber required for the new yacht and noted that some had been ruined by the carters cutting it to size to fit their vehicles. He found that Captain Taylor of Wapping had some very good planks. By August he was thinking about fittings, enquiring 'whether there be Holland duck in the stores at Deptford enough to make sails for the King's new yacht'. He would happily have her masts made at Woolwich. Three poop lanterns were to be provided, and he asked that 'Mr. Cockerell, the King's brazier at Chatham, may be spared to complete the chimneys of His Majesty's new yacht'. Plumbers were also needed to fix the pump. Ballast was a thorny issue again. Sixteen tons were needed, but stone was not suitable. 'If stones are used instead of shot for ballast of the King's new yacht, she will be damaged, for the quantity of stones required would make it needful to half-fill the cabins, and would make her run leeward.'

And when she was launched she proved to be a foot too light in the water, but the Ordnance Department was no more helpful. An offer of broken cannon from Chatham was rejected as they would take up too much room. The Ordnance officers reported:

> As to the demand of 10 tons of burr and musket shot, and six tons of great iron shot as ballast for the King's new yacht, so great a quantity has been lately issued that the stores are much exhausted, and some being daily required for its proper uses, it would prejudice the service to unfurnish the stores upon an account not in their cognizance, and for which they receive no allowance. Think the same may be better provided from the merchants by the Commissioners, who can defray the charge. Will supply any unserviceable shot in the stores.

She had been launched by 25 September after Pett requested a postponement until the next spring tide, so that her rigging could be more advanced. There was an incident in the river when a passing boat was confiscated for not striking their colours to them. It was restored, as the dockyard officers had no authority in the matter.

By October she had been given the name *Henrietta*, probably after the King's favourite sister, rather than his mother, with whom he was on bad terms. The carved work, presumably external, was certified to be 'well and sufficiently performed' by master carver John Ledman, who requested £50 'towards the payment of his workmen'. But the Navy Board was surprised 'to hear of so great a demand for carved work, when it was ordered to be forborne till trial had been made of her sailing.' Platerers were to work on the three poop lanterns, but it was November before the joiners' work was completed. Even so, in January 1664 the King 'commanded some alterations in the *Henrietta* yacht, turning the lockers into settle beds, and placing a small standing bed in the after cabin. Requests a warrant for that work.'

The smaller yachts

In 1661 the Dutch gave Charles another yacht, smaller than the *Mary* at 34ft long and 14ft broad, measured at 34 tons. The bezaan yacht (variously spelt in English) was a particular Dutch type, usually smaller than the states yacht and carrying a mainsail with a very small gaff at its head, not unlike a modern Bermuda-rigged yacht. The gift vessel, known simply as the *Bezan* or *Besane* in English, was fitted with leeboards in Dutch style and one was damaged in 1677.[13] According to Sir Anthony Deane, her hull was valued at £540, and £1141 fully rigged and stored, not much more than half the cost of the *Katherine*. In September 1661 Samuel Pepys, recorded, 'upon the Thames I saw the King's new pleasure-boat that is come now for the King to take pleasure in above bridge, and also two Gundaloes [gondolas] which are lately bought.' The use of such a small vessel above London Bridge was much more practicable, but in fact the *Bezan* would spend most of her career further downstream.

The type had a reputation for sailing close to the wind – in 1664 one was part of a group of captured Dutch ships, and she was the only one retained because 'she plies well to windward' and almost escaped.[14] Perhaps it was this quality which inspired the building of an English version. In October 1661 an estimate was produced for building a 'besane' yacht of 30 tons at a cost of £520. This was just after the race between the *Katherine* and the *Anne* in the Thames, in which ability to sail to windward seems to have been a major factor in victory. The new vessel, *Jemmy* or *Jamie*, named after the King's natural son, the Duke of Monmouth, was to be built by Christopher Pett at Woolwich, but was largely designed by Lord Brouncker, one of Pepys's colleagues in the naval administration. Pepys reported that she was designed 'according to new lines, which Mr Pett cries up mightily.'[15] She was to be built at Lambeth above London Bridge, and was decorated by the naval painter Isaac Walker at a cost of £160. She was smaller than the other yachts but bigger than the *Bezan*, 40ft on the keel and 16ft broad, measured at 54 tons, with a crew of ten men. Meanwhile, a smaller version, the *Charles*, was built by Pett at Woolwich.

Samuel Pepys sailed in 'the little Dutch Bezan' in September 1662. He testified to her qualities when she competed with the *Jemmy*, 'lately built by our virtuosos (my Lord Brouncker and others, with the help of Commissioner Pett also)'. She had presumably 'shot' or passed through the dangerous waters under London Bridge to get downriver from Lambeth. Pepys could not resist a dig at his professional rivals: 'before they got to Woolwich the Dutch beat them by half a mile; and I hear, this afternoon, that, in coming home it got above three miles; which all our people are glad of'.[16]

There is no sign that the *Bezan* was used by the royal family, and in 1670 she was at Gravesend transporting soldiers to ships. In 1673 she was used for a court martial, and later in the year she was used to pay and discharge victualling ships, but it was observed that she was beyond Sheerness and should be replaced for the winter. She was driven ashore in May 1674, and in October she was damaged in a collision with the *Christian* pink. In January 1676 she was used to return yacht crews who had taken the *Leopard* downriver, and she did the same for the *Woolwich* in April. Her defects were listed by the officers of Deptford in February 1677, and in September she was hit by the *Susan* flyboat of London, while at her moorings off Greenwich. In December one of her leeboards broke while she was sailing through the Hope, apparently in rough weather as she had two reefs in her sails.[17] There was a mutiny among pressed men in September 1678, and in June 1679 she was used to place buoys for Trinity House. In the same month, Sir Thomas Allin asked for her to take him back to London from the Nore, as he could not endure going by hackney coach.

Little more was heard of the *Charles* before she was given to the Board of Ordnance in 1668. Despite the rivalries, Pepys enjoyed sailing to Woolwich with Pett in the *Jemmy* in March 1663, 'being a fine day and a brave gale of wind'.[18] In August 1673 Commissioner Sir Richard Beach at Chatham had to deal with 'the [master?] Shipwright's complaint regarding the Jemmy yacht'.[19] In 1676/7 she was employed delivering masts and yards to a contractor at Woodbridge in Suffolk; she survived until 1721, but not as a royal yacht.

The *Monmouth* and *Merlin* were built in 1665/6 in private yards – Castle and Shish, both at Rotherhithe (or Jamaica Levels, as it was sometimes known), on the banks of the Thames between London and Deptford. The fact that they were ordered at the height of the Second Dutch War suggests that they might be intended for some military function, such as scouting or despatch carrying, in addition to their normal duties. They were typical in most respects, at just over 100 tons, but the *Merlin* was the only yacht to have a galliot rig, with a small mizzen mast and sail. In 1670 she was ordered to be painted for the voyage of 'Madame', the King's sister, the Duchess of Orleans, from Calais to Dover, though she seems to have travelled in the *Mary Rose* of 40 guns. The *Merlin* took part in various processions, such as the passage of Mary of Modena, also from Calais to Dover in 1673, and Charles's visit to the *Tiger* in 1681, but she was probably not carrying members of the royal family. In 1671 she was carrying Lady Temple from Holland when the Dutch fleet failed to give what was regarded as an adequate salute, a spurious pretext for the Third Dutch War. After that, she gained some distinction by being used by Captain Greenville Collins for his survey of most of Britain's inshore waters. The *Monmouth* went to the Irish Sea to serve the Lord Lieutenant of Ireland after the *Mary* was lost – she was considered more suitable than the *Merlin*, whose 'sharpness' made her 'less fit to lie on the ground' in the shallow waters around Holyhead;[20] then she served Collins for a time during part of his survey. Both were sold in 1698.

~3~
The Uses of the Yachts

Yachts were mostly kept in commission in peacetime, unlike the great bulk of warships. In January 1676 thirteen yachts were in service with a total crew of 210 men, an average of sixteen each, at a cost of £10,920. Nominally they were part of the 'Summer and Winter Guard' in the Downs in the North Sea, supplementing one fourth rate and two fifths, at a total cost of £14,040, but it is doubtful if the yachts did much guarding. In addition, the *Saudadoes* was allocated 'for the Queen's service' with a crew of fifty and another yacht, the *Mary* replacement, for Ireland with twenty men. More than 7.5 per cent of the peacetime manning budget was spent on yachts.[1] Charles could well have argued that the yachts mainly served diplomatic purposes, rather than his personal needs, but although they had some practical uses, the admirals persistently referred to them as 'the King's pleasure boats'.

Why did a perpetually cash-strapped monarchy spend so much on the yachts? Not for foreign travel: Charles never left England after his return in 1660, even to France to visit his beloved sister 'Minette' – Henrietta, Duchess of Orleans and sister-in-law to King Louis XIV. Perhaps it would have reminded him of his bitter foreign exile, or perhaps he feared that the fragile system of government would collapse in his absence. But the King was clearly fond of his new yachts, at least for the first year or so. On 21 May 1661 Samuel Pepys recorded 'we took barge again, and were overtaken by the King in his barge, he having been down the river today for pleasure to try it'.

Mary of Modena leaving Calais in November 1673 to join her husband James, Duke of York. (© National Maritime Museum, Greenwich, London, BHC0319)

William and Mary sail from Margate in November 1677. The yachts *Katherine* and *Mary* are in the foreground, with the *Charles*, *Anne*, *Portsmouth* and *Navy* further away, and a Dutch escort in the background. (© National Maritime Museum, Greenwich, London, PAH9364)

Yacht racing

On 1 October 1661, according to the diarist John Evelyn:

> I sailed this morning with his Majesty … on a wager between his other new pleasure boat [*Katherine*], built frigate-like, and one of the Duke of York's [*Anne*], the wager £100, the race from Greenwich to Gravesend and back. The King lost it going, the wind being contrary, but saved stakes in returning … his Majesty sometimes steering himself.[2]

This is often put forward as the origin of British yacht racing, but there is no sign that it was repeated. The River Thames is not ideal for yacht racing: it is narrow, winding, especially around what is now known as the Greenwich peninsula where the course has to be almost reversed, from north to south-southeast, and the river is strongly tidal. Evelyn implies that they made it to Gravesend and back in a single day, in which case they were extremely lucky with the winds, even if they were contrary for the voyage out – later log books often show comparable yachts taking several days to progress in one direction or the other, and anchoring to await favourable winds and tides. And on the trip in May witnessed by Pepys, it seems that the King left the yacht and was rowed back in his barge, rather than waiting for the wind or tide to change. Charles came to prefer the far more predictable sport of horse-racing.

Apart from that, the nearest thing to a yacht race was between several yachts in 1671, mainly to test the new *Cleveland*. There is no reason to doubt that Charles enjoyed the contest, but it was more of a test of a brand-new ship than a pure pleasure voyage. It seems that the race of October 1661 was a unique event, which attracted little notice at the time and did not set any trend, among the royal brothers or the population at large. Around 1670 another race failed to materialise, as the King told Samuel Pepys:

> … upon a certain occasion of one of the best sailing yachts of Holland coming over hither, and their ambassador van Bunninghen his might magnifying their quality of sailing, he offered van B to sail against her or any yacht of Holland with a yacht of his for £500 … but van B thought not fit to accept the challenge.[3]

The Duke of York's yachts

King Charles usually had several yachts available to him, to serve himself, his Queen (for a time), diplomats, his courtiers and his mistresses. His brother James, Duke of York, only had one at any given time, the *Anne* and then the new *Mary* from 1677. As a result of that, and because, unlike Charles, James had two legitimate daughters, James's yachts were kept far busier than Charles's, and the *Anne* and *Mary* only spent short periods at Deptford, except in the winter of 1675/6, when the *Anne* was used for various courts martial. They were the busiest of the royal yachts of the period, travelling as far afield as Danzig, though much of their work was diplomatic. In 1680 alone the *Mary* took James's wife, Mary of Modena, to Holland to avid a political crisis at home, then went to Spithead to pick up Governor Legge and Lord Hatton and take them to Jersey and Guernsey. She took the Dutch ambassador to Rotterdam and brought his successor back with his family. She carried Colonel Sydney, an envoy extraordinary, to The Hague, then fetched the French ambassador, the Marquis d'Agneau, from Calais. After that it was the Duchess of Monmouth and her children to Dieppe, Colonel Graham and a company of soldiers from Scarborough, then to Holland with the baggage of the Prince of Orange and to bring back an ambassador, Sir Robert Southwell, from Hamburg. Finally, she took Madame Loftus and her retinue to Dieppe before returning to Deptford for the winter.

Royal marriages

Yachts played their part in the arrival of present or future members of the royal family. In September 1660 the newly created Earl of Sandwich was sent to Holland to fetch the Princess Royal, sister of King Charles, Princess of Orange and widowed mother of the future King William III, to take her part in family affairs. They anchored in Westgate Bay on the north coast of Kent to be joined by the King and Duke of York: on 25 September, 'His Majesty, his Royal Highness the Princess Royal, the earl of Sandwich, with the rest of the nobility went out of the *Resolution* into his Majesty's pleasure-boat [presumably the *Mary*] in Hosewell Bay by St Margaret about 6 in the morning, and so went for London and as I [Sandwich] heard landed at London that night.'[4]

In the spring of 1662 Sandwich brought the future queen, Catherine of Braganza, from Portugal, in a fleet led by the *Charles*. As they passed Torbay, the Duke of York was waiting for them in the *Anne*, and came on board with several aristocrats and officials to welcome her. He did this several times over the next few days as the fleet proceeded slowly. 'The Duke kept in his yacht upon the quarter of the *Charles* and every day came on board to visit the Queen.' The fleet anchored at Spithead on 14 May and 'about 4 o'clock in the afternoon his Royal Highness took the Queen in the *Anne* yacht and sailed to the beach at Portsmouth next to the town and then her Majesty went into the barge wherein she was brought ashore.' She and Charles were married a week later.[5]

The yachts were involved in several other royal marriages, including the straightforward task of bringing Mary of Modena from Calais in 1673 as James's second wife. In the autumn of 1677 the *Mary* was sent to Holland to fetch William, Prince of Orange and Stadtholder of the United Provinces, for his marriage to Princess Mary, daughter of the Duke of York and third in line for the thrones of England and Scotland. The yacht arrived at Hellevoetsluis on 30 September and left with the prince on 8 October and landed him at Harwich the following day. The couple were duly married on 4 November

A yacht, probably the *Merlin*, sailing towards a three-decker, probably the P*rince* of 1670, by Willem Van de Velde the Elder. (Royal Collection Trust / © Her Majesty Queen Elizabeth II 2022, 400032)

and according to one report there was 'great rejoicing at the Prince of Orange's marriage, which was performed privately in the Duke's closet at St James's last Sunday night, and that night was five times consummated, as our young men tell us.'[6]

Soon it was time for the prince to return to his homeland with his new wife, and the Governor of Gravesend received orders 'to suffer two States men-of-war now riding in the river, appointed to attend the return of the Prince of Orange, to come up as high in the river and as near the Tower as shall be desired, in order to take in such of the goods of the said Prince and his train as are to be taken in by them'.[7] On the 14th the *Mary* received orders to fall down to Erith to receive their goods. Mary insisted in waiting to see her sister Anne through a bout of smallpox, and for her mother's birthday on the 15th. The prince and his tearful wife boarded at Erith on the 19th. They reached Sheerness two days later with the other yachts and the *Saudadoes* and *Greyhound*, but according to Captain Gunman, 'this afternoon it blew a very storm together with showers of hail, being most dismal weather'.

Princess Mary was still reluctant to leave her native land and the couple went ashore at Kingsferry where coaches were waiting to carry them to Canterbury. It still 'blew extreme hard' on Friday 23rd and boats were sent from Chatham bringing provisions for the yachts – even so, they still had to send men ashore to buy provisions suitable for the royals. The couple proceeded to Margate where they found Sir John Holmes and boarded his ship, the *Montague*. They tried to sail on the 26th, but it was reported, 'the wind coming N.E. in the night, they came back again into this road, and about 11 this morning came ashore to my house, where at present they are in very good health.' They boarded the *Mary* after some difficulty and sailed on the 28th. After a period of calm they passed the North Foreland and found an east-by-north breeze which allowed them to make 20 leagues or 60 miles in eight hours, a fast speed of 7½ knots. They began to sound the depths during the night and at four they saw the light at 'Sciven' (Scheveningen?), followed by the Goree light two hours later. They had to anchor in thick fog, and then approached the Naze, but they soon met 'whole islands of ice', which prevented them going into Rotterdam, so William required them to head for Terheyde, where they anchored in the afternoon of the 29th and put the royal party ashore.

James's younger daughter Anne was to be married to Prince George of Denmark and the *Mary* was sent to Gluckstadt to pick him up in the summer of 1683. Once there, Lord Churchill (as the future Duke of Marlborough then was) told them that the prince would not be ready for at least two weeks and sent them up the Elbe to Hamburg. He eventually arrived on 9 July with his father, the King of Denmark, but Captain Gunman still had to send a man ashore to inform Churchill that the winds were favourable and asking that he might 'hasten the Prince to embark'. He duly arrived at one that afternoon and Gunman set sail in company with the *Katherine* and made a fast passage

～ Royal visits ～

The yachts were often used for visits to the fleet. Early in 1662 the *Mary* was carrying messages to the fleet in the Downs.[8] In June 1661 the First Earl of Sandwich joined the fleet. 'Thursday. At about 11 of the clock I took barge at the Privy Stairs at Whitehall and boarded the *Mary* yacht at Deptford about 12 o'clock and so sailed for the Downs, where I arrived on board the *Royal James* on Friday in the evening.'[9]

After the Second Dutch War ended in 1667 there were occasional visits to the fleet. Commanders were not always given notice, and Sir Thomas Allin recorded:

> About one o'clock, I being at dinner aboard the *Royal Katherine*, news came they saw three or four yachts and colours flying which proved to be the standard. I hastened away and went aboard the *Anne*, where the King, his Royal Highness, Prince Rupert and many lords and gentlemen were, and went aboard the *Monmouth* and stayed there till 3 o'clock and then went aboard the *Katherine*, thence to Deal castle, and from thence to the *Resolution*, then to the *Montagu* and so aboard the *Henrietta* yacht. I received my orders to sail next day for Portsmouth.[10]

Charles sometimes visited parts of his kingdom by sea, even when land transport would have seemed more convenient. They were more than pleasure cruises, as they were invariably to places of strategic importance, where the King looked at fortifications as well as ships old and new, and dockyard facilities. He began in August 1662: 'About 5 o'clock the King came into the Downs in the Dutch pleasure boat [*Mary*] with Prince Rupert, the Duke of Buckingham and diverse persons of quality. They went ashore and lodged with Captain Titus in Deal Castle.' On the 24th the King dined at nearby Sandown Castle, took dinner in Admiral Sir Thomas Allin's pinnace and 'went aboard the pleasure boat and was wet coming shore.' Next day he boarded the pinnace again to go out to the yacht, and back to London.[11]

During wartime such trips were suspended and replaced by visits to the fleet. The King went slightly further afield in 1668, to the strategically important port of Harwich. The yard was largely on a care and maintenance basis since the end of the Second Dutch War the previous year. It was run by Captain Silas Taylor, the Storekeeper, a grander office than it might sound to modern ears. Having short notice of the visit, Taylor lamented that he had 'neither boat nor barge fit to receive his

Portsmouth c1675, by Hendrick Danckerts. It shows a yacht leaving the harbour, with the fortifications including the Round Tower to the right, Gosport on the other side of the entrance to the left. It was commissioned by Charles II but it does not show one of his visits as the yacht does not fly his standard. (Royal Collection Trust / © Her Majesty Queen Elizabeth II 2022, 405156)

Majesty', and had a stage or pier constructed into the water. The King duly arrived in the *Henrietta* and *Anne* yachts at three in the afternoon of 3 October, with his usual cohort of royalty and nobility, including his brother, his illegitimate son, the Duke of Monmouth, and two more dukes. He viewed the shipyard where the classic 70-gun ships *Resolution* and *Rupert* had been launched, as well as the cranes, one of which survives today. He spent the night on board the *Henrietta* and landed at six next morning for the main attraction – the building of Landguard Fort by Charles's chief engineer Sir Bernard de Gomme: 'he went on foot out to the town, viewing all the places in relation to fortifications, and examining some drafts offered by Sir Bernard, which he rectified in the field at two or three stations, with his own hand, by a black lead pen and ruler.' He sailed along the coast in the yachts to Aldborough that evening and next day he rode to Ipswich to begin the voyage home by land.[12]

In 1671 the King travelled by road from Windsor to Portsmouth. After taking part in a trial between several of his yachts (see above), he sailed west with a small group of ships and yachts – 'we suppose for Plymouth', as Hugh Salesbury, the mayor of Portsmouth, put it. For it was not just the weather which made Charles's trips unpredictable, the King's mercurial nature was another factor, as this trip would show. At Plymouth the Earl of Bath had very little advanced warning, but he 'made all possible preparations for his reception the time would allow'. He went out to sea to meet the visitors and they came ashore on the 17th to be 'welcomed with a peal of all the ordnance from the citadel, St Nicholas Island, and Mount Batten, after which all the ships and merchantmen in the Sound fired all their guns' – for royal visits were noisy affairs. Charles was impressed with the building of the citadel. Plymouth was not yet a naval port, but next morning the King viewed several sites, including the Hamoaze, where his successors would establish the Royal Dockyard. Charles remarked that Plymouth was 'that most excellent harbour beyond any other in his dominions', but had a sudden urge to sail, much to the surprise of the mayor and magistrates, who did not have 'time to kiss his hands at his departure.' It was even more of a shock to the dignitaries of Devon and Cornwall, which were 'more populous of gentry than in any counties of England'. They flocked to Plymouth, but had no time to meet the King or kiss his hands. And Pepys recorded, '[o]ur stay at Plymouth was so short'. The King embarked in the *Cleveland*, now established as his favourite yacht, and set sail at nine in the evening.

If the King had sailed because of some premonition of favourable winds, he was quickly disappointed. Next day the small fleet anchored in Torbay and it was noted: 'The wind being contrary will make his passage long.' On the 20th they sailed close to the shore of Lyme Bay, where it was observed, 'the wind being strong NE and E, they could not get about Portland, but next morning tacked to the westward again.' They went to Dartmouth and the King ordered his coaches to come from Portsmouth to Salisbury as he intended to return by land.[13]

On 12 June 1675, after the Third Dutch War, it was recorded that the King was 'very intent' on going to Portsmouth to see the launching of a new first rate ship, the *Royal James*, planned for the high spring tide. The yachts *Saudadoes*, *Portsmouth*, *Anne*, *Katherine* and *Richmond* were ordered to be got ready, followed by the *Monmouth* to carry the duke of that name – although the King intended to travel in the sixth rate ship *Greyhound*, which was cleaned and fitted at Deptford alongside the 18-gun *Lark*. They were ordered to go downriver to Gravesend to meet the King, though Samuel Pepys intimated that he would travel by land. He warned his brother-in law, Balthasar St Michel, the muster-master at Deal, to be ready to expect the yachts as they passed through the Downs.

Charles duly left Gravesend on Saturday 26th, but the wind 'came round the compass', and they anchored between the Cant and Oaze Edge Buoys. By four in the morning the King was up and noticed a favourable wind. He fired a gun for the fleet to sail. The wind carried them to the Red Sand Buoy in the middle of the Thames Estuary, off the entrance to the Swale, where they waited for the tide. They passed over the flats at nine, and rounded the corner at North Foreland by midday. The King was 'resolved to ply towards Portsmouth' and was 'making all sail they can away.' They sailed south past the Downs off the east coast of Kent, where they were joined by several other naval ships. About eighty boats set out from Deal to welcome them, each flying a jack and ensign, 'to present our obedience to him, which he was pleased to accept of.' Moreover 'flags were hung up in Deal town'. They sailed round the South Foreland and past Dover. They had almost rounded the promontory of Dungeness, but the wind was now against them and they had to turn back to the Downs, where they anchored four miles equidistant from Ramsgate and Deal. They were now in a 'short scurvy sea' with wind from the south-southeast, a 'topsail gale', in which all sails except the topsails would be taken in on a ship at sea. Boats could not put out from the shore, and the landsmen were distressed.

Finally at three in the morning of the 29th, the wind turned northeast and the voyage progressed. They were off Dungeness by seven and Beachy Head by eleven, though they were already too late. The officials at Portsmouth expected the King hourly, but they could not afford to miss a favourable tide. The *Royal James* duly took to the water at eleven that morning. Samuel Pepys wrote, 'the new ship was very happily launched, but without any tidings of his Majesty'. Meanwhile, at four in the afternoon the King's flotilla sighted the Isle of Wight, but the weather turned hazy. Fearing running aground on the island, they turned more southerly. Pepys's report includes the phrases 'His Majesty set sail' and 'The King then plying to windward', which might or might not imply he was making the decisions personally. In any case, according to Sir John Werden, the Duke of York's secretary:

> the weather proving very thick, perhaps too our compasses being disordered with the violence of the sea, we so far outran or mistook our course that late at night we found ourselves to the westward of the end of the Isle of Wight and then we fell to ply to windward in very stormy and dark weather and thus lost company and sight of his Majesty in the *Greyhound*.

Werden was not enjoying the 'very tedious voyage', and the duke's yacht *Anne* was 'a very bad seaboat'. At one in the morning the *Anne* anchored in Sandown Bay in the south of the island, while the *Monmouth* with the eponymous duke on board plied about all night without anchoring, then joined the *Anne*. When the weather cleared in the morning they saw the *Greyhound* twelve miles westward, apparently intending to enter the Solent from the west and go to Freshwater Bay. Instead, on the morning of 1 June, she anchored off Dunnose, just east of the southern tip of the island.

The weather was still bad, but eventually the King was landed in the *Greyhound*'s shallop, to be met by the governor, Sir Robert Holmes. The buccaneering seaman was wealthy from prize money and he entertained the King at Yarmouth at the western end of the island, before sending him off for a belated visit to Portsmouth, where he knighted Sir Richard Haddock, and the shipwrights Sir John Tippets and Sir Anthony Deane.

> In seeing one of the yachts, built here for the French King at Versailles, drawn on a cradle placed on four wheels at least 200 yards to the seaside, where it was lifted up with tackle and other engines (though it weighed at least 42 tons) and let down gently at once on the ooze, where the tide came in to it, and this afternoon we have seen it sail about with great applause.

According to Pepys it seemed 'to outdo anything that yet ever swam.'

The *Katherine* yacht was still missing, but the King was not deterred:

> This stormy voyage has not at all discouraged his Majesty from the sea, and all he can be persuaded to is only to change his ship and return in the *Harwich*, a good third rate frigate, but he will by no

means harken to any proposition of returning by land.

Pepys was relieved, for he had little confidence in the King's judgement in these conditions, or the competence of the captain of the *Greyhound*. The change would lead to 'the lessening though not wholly removing the apprehensions we were lately under from the two great adventures he was then running without other security on board him but his own seamanship, and poor Clements.' George Legge agreed that the *Harwich* was 'the best man-of-war with him, and I think, the safest.' The King arrived in the Downs at ten in the evening of the 4th, with only the *Saudadoes* and two small ships in company.

This was the peak of Sir Anthony Deane's shipbuilding career. His *Greyhound* and *Lark* had been specially chosen for the voyage. According to Werden, his new ship, the *Royal James*, was 'by all acknowledged to be the most complete piece his Majesty has in all his navy.' He had built the *Harwich*, in which the King was to return, and according to Pepys (who was not unbiased), it 'carries the bell of the whole fleet.' And his yachts for the King of France were highly acclaimed, though they were the harbingers of trouble.

Naval affairs took up a good deal of Charles's attention during 1677, when he and Pepys steered a bill through Parliament to augment the navy – Parliament agreed to provide £600,000 to build thirty large new ships.[14] In August Charles decided to visit Portsmouth, using his yachts this time, with himself in the new *Katherine* which had replaced the ship lost to the Dutch. At seven in the morning of the 7th a fleet of seven men-of-war and eight yachts passed Dover, but again it was driven back to the Downs, 'the flood tide coming and the wind blowing very hard with rain'. They sailed between four and five the next morning with a southeasterly wind. The wind veered more westerly, but it was hoped they would get to Rye or Hastings that night. By midday on the 9th they were six leagues southeast of the Isle of Wight, when the King changed his mind. Captain Clements, still in the *Greyhound*, was sent into Portsmouth with the message that with the favourable wind, His Majesty intended to go on to Plymouth. The fleet anchored overnight in St Helen's Road at the eastern end of the Isle of Wight. The *Kitchen* yacht ran aground and was damaged, but the fleet sailed again, despite warning of bad weather, and were out of contact for several days.

Everything was prepared in Plymouth, guided by the Earl and Countess of Bath, but there was serious worry about the King being out in such bad weather. When a messenger arrived overland, '[h]is good news raised us all from the dead'. The King came in with the tide at three in the morning of the 17th, with only two other yachts in company. They anchored in the Sound and three hours later they used the tide to carry them into the harbour, being saluted by St Nicholas Island, Mount Batten and the Citadel. There were no ships to launch, but Charles went round the new fort and was 'much satisfied with the works'. Besides its military strength, it had the most elaborately decorated entrance in the country, but it was still unfinished, despite the expenditure of more than £20,000 up to 1675, and it was not fully complete for more than a hundred years. He went shooting in the Hamoaze and at Saltash, and was received with much enthusiasm by the people – he even planned to visit there for a month or two each year and left De Gomme to design a house for him.

∽ Passengers ∽

In February 1663 Thomas White pointed out, 'The number of passengers between Calais and Dover is much lessened by weather and by the King's pleasure yachts, which are constantly conveying lords to and fro, and carry all by reason of their gallant accommodation.'[15] The career of Captain William Faseby (or Fasby) illustrates much of this work. After commanding the *Anne*, *Kitchen*, *Katherine* and the *Anne* again, he was appointed to the new *Cleveland* in June 1671. In September 1673 he was ordered to Rye where he would pick up the Earl of Sandwich 'with his company, goods and servants', and take them to Dieppe before returning to Greenwich. In July 1674 he was to go to Rye again to pick up Lady Henrietta Hyde, and to take her 'unto such port in France as she shall direct'. He was to wait four days, then bring her back. In January 1675, with the first opportunity of wind and weather, he was to sail to Calais to wait for Lord Douglas, and then 'bring him over unto such port in England as he shall desire.' In July he was ordered to escort Deane's yachts for the King of France across the Channel to Le Havre. He picked up Lady Goring at Dieppe in September. He was transferred to the *Charles* yacht a month later under a commission countersigned by 'Mr Pipps'. In June 1676 he was ordered to take the French Marquise de Bethune to Danzig, with strict instructions not to strike his topsail to Kronenburg Castle at the entrance to the Baltic. In May 1677 he was to wait at Dieppe for four days for Lady Goring and take her wherever she wanted in England. Things went wrong in the spring of 1678, when a letter was suppressed by the master of a Dover packet boat, ordering Faseby to fetch a Mr Mackland from Calais. Instead, he 'was driven to take his passage upon the packet-boat, and thereby hurt and pillaged by a pirate that met him at sea.'[16] In July 1678 Faseby was to receive on board the Earl of Feversham and take him to Brielle in Holland, or 'such port in Holland or Flanders as he shall desire.' After the loss of the *Charles* off the coast of Holland in 1678, Faseby left the yachts to become captain of the 70-gun *Kent*.[17]

Some of the voyages were diplomatic. In January 1674 the King approved 'appointing the *Bonadventure* to convoy over the *Cleveland* yacht with the French ambassador.'[18] In September 1676 Captain Day was ordered, 'To receive on board the *Bezan* yacht Sir Robert Robinson, with what goods either of his own, or the Lord Dungan, or the Portugal envoy he shall direct, and to carry him down to the *Assurance* at the buoy of the Nore'.[19] In February 1682 there was a Treasury warrant to the Customs Commissioners 'to pass, customs free, the goods of the Russian ambassador, his son, secretaries and retinue, on the *Mary* yacht.'[20]

The yachts often carried valuable goods for royalty, nobility and others. In April 1681 the *Henrietta* brought 'pictures, pourcellin and Boucaros, writings and books of history, eight pieces of old tapestry hangings, 10 pieces of new tapestry hangings, pictures with their frames, painted screens and pictures, etc' from Brussels for the Duke of York. In June Henry Guy was given permission by the Customs Commissioners 'to deliver at St. James's House three boxes directed to her Royal Highness Lady Anne [said boxes being now] aboard the *Mary* and six chests for the Duke of York directed to Sir Allen Apsley in St. James's Square.' In April 1682 the same Commissioners allowed 'the import, Customs free, of a box of spices, on the *Katherine* yacht, Capt. Davis commander, from Holland, directed to the Earl of Conway, being a present to him.' And in November 1683 they were ordered 'to open at the Marquis of Richelieu's house in Duke Street, St. James's, a large trunk lately arrived on the *Fubbs* yacht for him.' In June 1675 the *Portsmouth* and *Anne* carried 'several vestments, habits, scenes and other necessaries belonging to the Italian comedians [ie actors in comedy plays] lately brought from France'.[21] They were not the only actors so favoured. In May 1684 there was mention of '65 trunks or packs of old clothes belonging to the French Players of the Prince of Orange's Company and coming at the King's desire from Holland in one of the King's yachts which he has sent for them.'

A passenger could expect to incur some expenses in a trip, though they were paid from Treasury funds in the case of official voyages. In April to July 1683 Sir Leoline Jenkins, the Secretary of State, paid £6 9s 0d 'to the Captain of the yacht who carried me to Dieppe', six guineas to the master and seamen, £3 12s 0d for a barge to carry his goods on board, £10 10s to the *Kitchen* yacht for carrying some of his goods and servants, and a guinea 'to the boat's crew who landed me from the yacht which could not come into the road.'[22]

This anonymous painting could be depicting one of Charles's visits to a naval installation. Yachts are seen to the left and right with a small warship (perhaps the *Greyhound*) flying the royal standard to signify the King's presence in the right centre background, with a larger French warship beyond. (© *National maritime Museum, Greenwich, London, BHC0835*)

4
The Later Yachts

The government's financial situation was increasingly difficult after the Second Dutch War of 1664–67, and the King's popularity diminished after the disasters of the Great Plague, the Great Fire of London and the Dutch raid on the Medway. That did not prevent the building of yet more yachts, not all of which were replacements for those lost.

The *Kitchen*

In 1670 Charles perhaps remembered the yachts which had fed them well during his glorious passage through the Dutch waterways ten years earlier, supplying provisions of 'so prodigious quantity of all sort of meats, of foul or sweet meats, of wine'. On 19 April Samuel Pepys asked the Navy Board, of which he was a member, 'for money to build a kitchen yacht'. The idea was not entirely new: in 1663 William Faseby was master of the hoy *Kitchen* and it was suggested that he should meet the Duke of York at Southampton 'as there will be want of the kitchen there.'[1] But this was a purpose-built vessel, not one converted from a small Thames river craft. They ordered it to be considered, but things moved fast after that. There was an Order in Council to give it the highest authority short of an Act of Parliament, and the work was given to Mr Castle in his yard at Rotherhithe just above Deptford, rather than in a royal dockyard like most royal yachts. Work proceeded fast: on 14 June it was '[o]rdered that £800 be provided for Mr. Castle for building the *Kitchen* yacht.' Three weeks later there was a 'Treasury order on the fee farms for £922 to the Navy Treasurers for building the *Kitchen* yacht.' She was in commission by 20 May.

Anthony Deane recorded a total price of £830, including rigging and stores. Her hull only cost £400, a price of £5.26 per ton, compared with £7 for the standard yachts, and considerably cheaper than even the *Merlin*, and is probably explained by the much lower cost of decorations. She had four iron guns, the only yacht not to have brass guns in 1670, and they cost £9 18s compared with more than £400 for the standard yachts. She had a standard crew of twelve, but it is not clear whether that included the cooks. A Van de Velde drawing shows her with comparatively simple decorations, square rather than round wreaths round the gunports, only lightly carved.

Despite her name, her regular complement only included one cook – Daniel Colman in the early 1670s – who presumably served the ship's company, with royal cooks being brought on board when needed. She was depicted carrying out her main function during the King's visit to the fleet in July 1673. Van de Velde captioned his drawing: 'This incident shows the food being brought over on Sunday the 16th June [continental style] 1673.' Boats are rowing the food from the *Kitchen* to the *Cleveland*, probably in Sea Reach on the Thames. Both yachts are under way, presumably sailing quite slowly or the boats would not be able to keep up. The exercise was repeated during the King's second visit in July, though this time the yachts are shown at anchor downstream from Woolwich – 'Midday, the King's food is handed over and lunch is served.'[2]

Such an operation was only possible in sheltered waters such as the Thames or Solent and was probably quite rare. At other times the *Kitchen* was to be found ferrying soldiers from Hull in August 1670 and carrying seamen and their clothes to Portsmouth in the following month. Lieutenant Thomas Andrews asked the Navy Board 'that the *Kitchen* may be ordered to take down 30 volunteers, entered on board the *Princess*, under Sir Wm. Jennings, they being in town, and complaining of being kept out of victuals and pay.'[3] She retained her cooking facilities until June 1692, when William III's Admiralty ordered the Navy Board, 'to cause the kitchen

Charles's mistress Louise de Keroualle, Duchess of Portsmouth. Two yachts, the *Portsmouth* and *Fubbs*, were named after her. (Royal Collection Trust / © Her Majesty Queen Elizabeth II 2022, 405882)

in their Majesty's *Kitchen* yacht to be forthwith taken away, and the said yacht fitted with accommodation in like manner as the other yachts'.[4]

Deane's yachts

In 1669/70 Pepys's friend and protégé Anthony Deane, then Master Shipwright at Portsmouth, built a small yacht for the use of the Queen – he had already built the tiny *Fanfan* of 33 tons, which for some reason was registered as a sixth rate rather than a yacht. Little is known about the new yacht except for her dimensions – 51ft 6in on the keel, 17ft 6in broad, 8ft deep and measured at 83 tons, with eight guns, which must have been very small. On completion she was sent to the Thames where it was reported in April: 'The Queen was entertained this day by the Duchess of York at Deptford, where she went on board her ship, gave it a Portuguese name, and fired a gun.'[5] The name in question was *Saudadoes*, which was often misspelt. It is untranslatable but indicates a bitter-sweet longing for something unobtainable and perhaps represented her desire for her homeland. The yacht was used above London Bridge, as reported by one witness. 'Her Majesty gives life to all by frequent divertisements upon the river in her new vessel the *Sodalis*. They undertake long voyages, and, falling short of provisions, victual themselves sometimes at Vauxhall and sometimes at Lambeth Palace.'[6] This phase did not last long: by July 1673 she was in the hands of Jonas Shish, the master shipwright at Deptford, who appeared to criticise her design and aspired to 'better her qualities or render her more fit'.[7] What emerged was classed as a rebuild and was certainly far more than a repair: she was more than doubled in size to 180 tons with sixteen guns and seventy-five men. She was now classed as a sixth rate warship, but the King showed special concern over her gilding: it was to match that of the *Greyhound* which had already been used to transport the King, and she was to be furnished with a set of silk pennants. In June 1674 the Navy Board was ordered to 'grave all the King's yachts, and also the *Saudadoes* and *Greyhound*.'[8] She was still listed as being 'for the Queen's service', though there is no sign that Her Majesty was on board her after the rebuild. Instead, she made several voyages to Portugal, maintaining a tenuous link with her homeland. After 1688 she served as a small warship.

Deane's first fully-fledged royal yacht was the 107-ton *Cleveland*, named after the duchy of Charles's mistress Barbara Palmer. The building at Portsmouth produced little surviving correspondence, partly because this was a rather dry period for naval records, but she was launched in 1671 and soon proved a good sailer. Soon after her launch she was tested against the others and it was reported, 'Friday morning he sailed to try all the yachts, and from thence to Newport in the Isle of Wight is about nine leagues. They all sailed with contrary winds, and in about four hours' time got thither, in which time

The *Kitchen* yacht showing a plainer and simpler style of decoration. (© National Maritime Museum, Greenwich, London, PAH3929)

the *Cleveland* got to an anchor before any of the rest.' She was soon established as Charles's favourite.

In 1673 she was followed by the *Navy* yacht of 74 tons, whose name might suggest that she was intended for the service of the Navy Board, but there is no sign that she carried the said commissioners about their duties. She spent the Third Dutch War ferrying troops and seamen, carrying up to seventy-five at a time, and briefly as personal transport for Admiral Sir Thomas Allin. Later she served the Governor of Guernsey by taking dignitaries, troops and supplies between there and Southampton. After the Revolution of 1688 she was stationed in the Irish Sea for a time.

Deane's next yacht, the 120-ton *Charles*, was a replacement for the one which had been transferred to the Ordnance Office in 1668. Her financing was complex, and in January 1675 an Admiralty Board meeting was told:

> … the *Charles* yacht, lately built by Sir Anthony Deane and put to the charge of the officers of the navy, and entered into the list of the Navy Royal, was built by particular contract made between his Majesty and him, and having fully satisfied himself in the said Sir Anthony Deane's full performance of the said contract on his part, and paid out of the privy purse to the said Sir Anthony £1000 in part of the price of the said yacht, which his Majesty was pleased to declare to be £1500; it was his pleasure that the remaining £500 should be made good to him out of the Treasury of the Navy …[9]

However, the King clearly did not intend to pay for crewing and maintenance. 'His Majesty was pleased to declare his pleasure that the *Charles* yacht, newly built by Sir Anthony Deane, be taken into the list and charge of the Royal Navy, and forthwith fitted in all respects of his service therein.'[10] The charges included an estimated £300 for gilding.[11] But her life was short – she was wrecked on the Dutch coast in 1678.

Other yachts

Another type of yacht was beginning to emerge, not for personal use by the King and his entourage, but for some of his officials. These included the Lord Lieutenant of Ireland, who had the original *Mary* at his disposal until her loss, then the *Monmouth*, and the Governor of Guernsey, who had the *Navy* yacht, among others. The *Queenborough*, built at Chatham in 1671, was mainly used to service the isolated dockyard at Sheerness. The *Isle of Wight*, built at Portsmouth in 1673, was at the disposal of the governor of that island. The *Deal*, built at Woolwich in 1673, was under the orders of the Commissioner at Portsmouth until 1678, when she was transferred to the officers at the Downs, including Pepys's brother-in-law, Balthasar St Michel, and Admiral Sir Richard Rooth. Mostly she carried stores and rigging to ships at anchor.

Portsmouth 1674

In September 1673 Phineas Pett, assistant master shipwright at Woolwich, and one of the ubiquitous dockyard family, was summoned by the King to bring a plan for a new yacht to replace the *Henrietta*. The King approved it, but not for the last time there was confusion about the ordering procedure – Pett had only received formal orders to estimate the cost, not to begin the actual building. The Lord Treasurer was present and Pett was given an order to find timber in Epping Forest. The King then came to tell him that Johnson the shipbuilders of Blackwall had compass timber suitable for forming the frames of the new yacht. He duly went to the yard and measured some timber to be bought at five shillings per load ready money – for like all contractors Johnson was wary about delayed

Sir Anthony Deane, Samuel Pepys's favourite naval architect and the designer of yachts for the English and French kings, painted by John Greenhill around 1670. (© National Maritime Museum, Greenwich, London, BHC2645)

payment by the Crown. Pett was beginning work by 10 September, but a month later the shortage of timber was affecting the whole yard and workmen were being paid for doing nothing.

Lead was a problem as usual, for both plumbing and ballast; on 23 September Pett asked for old lead from Deptford and Woolwich – Captain Richard Boult cast it in his house where he had old lead and moulds. By January 1674 the plumber was ordered to make lead work for the yacht. By the beginning of October Pett was beginning to think about the carvings of the new yacht. In this case it was labour, rather than timber, which delayed the work, and in December 1673 Pett asked for permission to press twenty carvers to work under Mr Fletcher the master carver, over the whole yard. The yacht was launched early in 1674 and in April she was moved to Gravesend to prepare for the King's visit. There was a setback at the end of the month when her mast gave way in a strong gale of wind. Pett did a temporary repair by 'fishing' and 'wolding' it, but recommended a new mast of Riga fir, despite the expense to the King.[12]

Katherine 1674

By 7 October 1673 Commissioner Sir Richard Beach at Chatham had received verbal orders to begin another yacht, though he questioned on what authority it was to be done. Again there were problems with timber supply – Phineas Pett pointed out that they needed two rising floor timbers of rather specialised shape, which were not available in the yard or by contract. However, he found some in a timber yard at nearby Strood and asked for directions. By mid November the new yacht was taking up all the timber in hand and was 'a great hindrance to progress on other work'; in December he asked for fifty more shipwrights, because the yacht was delaying work; in January 1674 Beach complained that even work on the yacht would soon be at a standstill without extra resources. By November they were arranging for the supply of rigging blocks and anchors, the latter to be made by John Ruffhead, an anchorsmith of Deptford. Again old lead was recast for the yacht, but in November Pett complained that there was not enough in store. Yachts were built by the regular dockyard workers and timber mostly came by established dealers who were used to naval ways, but carving was a specialist task usually performed by contractors, who needed payment.

In February 1674 Fletcher demanded an imprest or advance for carved work to proceed. On 8 April Beach was thinking about the launch, asking whether the King would attend and whether there was to be a dinner in Hill House above the dockyard. That would require some urgency – Pett asked for the joiners fitting out the interior to be allowed to work 'tides' or overtime, and even to work through the Easter holidays.

The changing balance

In 1673 James, Duke of York, was banned from holding office in the navy by the Test Act, which was intended to root out Roman Catholics. In February 1677 King Charles II reopened Parliament after a recess of fifteen months and told the members, 'I desire you to consider the building of more ships, and how much all our safeties are concerned in it' – one of the first references to the Royal Navy as a national instrument, rather than the King's personal property. One speaker took the King's point a little further – they would be 'Parliament's ships' and 'what we give, we give not to the King, but for our own defence.' This was going too far and he was rebuked, but his point was valid. Parliament agreed to provide £600,000 to build thirty new ships.[13] And in 1679 Charles was forced to dismiss a stable and experienced Board of Admiralty and appoint one from the parliamentary opposition. Pepys commented that a board containing great knowledge and experience was replaced by those who were 'wholly ignorant thereof'. It would not be surprising if Charles was finding that control of his beloved navy was slipping away from him. But on the other hand his personal revenues, granted to him for life by Parliament out of duties on beer and wine, were growing with increased national prosperity. He was never clear of debt, but now he had a little room for manoeuvre. It is possible that he saw his growing fleet of royal yachts as a kind of private navy, and he certainly did his best to set his private stamp on them.

Perhaps that is why he paid so much attention to yacht design, a field where he could have an almost free hand, and in which he considered himself an expert. Samuel Pepys, now out of office, recorded, 'He pleases himself mightily with a design he is now going upon, of building a yacht the size of the *Katherine,* that shall go with a ketch's sail.' Later Pepys saw 'the original draught and dimensions drawn by the King's own hand of the yacht building by Sir Phin[eas] Pett at Greenwich, 1682'.[14] In fact the resulting vessel was rather different from the *Katherine*: according to Pepys's own figures she would be 7ft longer, 4in narrower, drawing one inch less water and measured at 148 tons, thirteen more than the *Katherine*. And it is not likely that the King drew the lines himself, however much he may have contributed to the design.

Charlotte 1677

Having partly financed the *Charles* from his own money in 1675, the King was using his private funds again in January 1677, when he told an Admiralty meeting, 'that he has given order for £500 to be presently advanced from his privy purse to the Treasurer of the Navy towards the building of a new yacht which he designs to have built by Mr Pett at Woolwich', and on 22 March he promised that he would 'give order to Mr

A yacht believed to be the ketch-rigged *Isabella*, drying sails and flying the Royal Standard at the masthead. Fishermen in the left foreground and major warships in the background. (© National Maritime Museum, Greenwich, London, BHC 3422)

May to pay it presently.'[15] On 30 June the board was notified of 'A bill of charge amounting to £46.10s by Sir Samuel Morland [the mathematician and inventor] for the brass frames and roller for his Majesty's new yacht at Woolwich'.[16] It was an exception to the 'standing rule in the navy that whatever charge in any of the said works shall be proceeded upon by a master shipwright, or his order, contrary to, without, or beyond the direction of the officers of the navy, … shall be made good to his Majesty out of the wages and salary of the said master-shipwright.' She was afloat by mid June 1676, when the Admiralty ordered, 'that the *Charlotte* yacht be forthwith caulked and graved'.[17]

Mary

The new *Mary* yacht, built at Chatham in 1677, was the largest yet at 155 tons. She had all the usual problems. There was confusion again and in February Commissioner Beach asked 'what orders have been sent about building the new yacht'. Pett had already selected two pieces for the keel at Deptford, but found them to be too short. The ironwork was done by Ruffhead again, following the precedent of the *Katherine*. Again the carvings were slow, in July Beach wanted 'orders for hastening of the new yacht and that she may be launched before her carved works are completed.' In November he complained that the shipwright was extravagant with the new yacht's boat, while Walker the painter pointed out that gilding could not be done in frosty weather. Her first task was to take the newly married William and Mary across the North Sea

Henrietta 1679

The building of a new *Henrietta* was perhaps a little more regular, in that Charles consulted his officials. According to the Admiralty Minutes of 55 February 1679, 'His Majesty is also pleased to declare his intention of having another yacht built for Captain Faseby [who had been in the wrecked *Charles*] by Mr Thomas Shish at Woolwich, to be of the same bigness as the *Charlotte*.' A week later the board approved an estimate with orders 'to limit the charge of her carving and gilding to that of the *Cleveland* and *Charles* yachts.'[18] Moreover, there were plenty of precedents which were cited during the building, though none of that made the financial situation any easier.

By the beginning of March Shish proposed making a new building slip for the yacht 'below the single dock in the yard', and this was done. He found some suitable compass timber and knees belonging to the landowner Sir Charles Bickerstaff, but he was still asking for more timber on 7 April, plus plank and elm timber for carving by the end of the month. He was still asking a month later, though the King wanted her built as soon as possible. In July the carver needed an order to make rails, for '[t]he King wants the yacht adorned like the *Charlotte* yacht.' There were the perennial problems with lead ballast; old lead was received from Chatham, but Shish wanted 10 tons more. In September the carvings were held up by lack of money and Mr Helby refused to proceed without it. Meanwhile, the painter needed £200 for gold and paint. By 3 October the hull was complete apart from carving and painting, which Shish tried to hasten. Isaac Walker the gilder demanded £300, which the Navy Board refused. Eventually Shish asked that Walker's foreman should only gild the outboard works, while the painter would 'proceed with the outside and leave the cabins alone.' On 16 December he excused himself from attending a Navy Board meeting 'because of launching the new yacht'. She was ready the following day, except for copper sheets and funnels, and scheduled for the next high spring tide on 23 December.[19]

The *Fubbs* and the *Isabella*

In 1680 another Bezan yacht was ordered from Phineas Pett at Chatham Dockyard. Named the *Isabella*, she was to be 46ft long and 16ft broad. In July Chatham asked for a pair of bilgeways to be made and the use of two old masts from Sheerness for launching gear, and the King was expected to attend on the 17th. But apparently the ship was not a success. In August Phineas Pett sent a list of necessary alterations and additions.[20] By October 1682 there were plans to convert her to a ketch but in 1683 she was sold to Pett, leaving the name free for another yacht.

The new *Fubbs* yacht was to be built at Greenwich rather than one of the royal dockyards, which perhaps gave Charles

A Van de Velde drawing showing the King's visit to HMS *Tiger* off Gravesend in 1681, showing the *Tiger* in the centre with yachts to the left and right and boats ferrying the King and his courtiers to her. (Collection Museum Boijmans Van Beuningen, Rotterdam. Photographer: Bob Goedewaagen, MB 1866/T 178)

more control. She was to be 63ft long and 21ft broad, exactly one third. It was not unusual for a naval official such as Pett to undertake private work, but he still needed the help of the dockyards for fitting; in May 1682 he asked for masts and yards to be made in Deptford Dockyard, rigging 'of fine yard (as the *Mary* yacht was) at the King's ropeyard at Chatham of such dimensions as I shall advise'; and blocks and pumps by the master smith at Chatham. Carving would be performed by 'young Mr Fletcher' to an estimate of £300, the painting and gilding by Walker of Deptford for £350 and the sails by a Mr Edmondson 'with His Majesty's canvas of such dimensions and sorts of cloth as I shall advise.' He promised to 'give timely notice' of any other work needed.[21]

On 26 May Mathias Fletcher reported that he had started on the carvings but needed money to buy necessities, while Chatham and Deptford yards were sending old lead for ballast. Thomas Wilshaw of Deptford yard was ordering canvas to Pett's directions and a mast was ordered from William Wood. By 7 August arrangements were being made for the launch on the 23rd, but Walker the painter had to be ordered to hurry up. The Deptford officers were 'to send to Greenwich the ways, ribbands, bilgeways and piece of wood for mast for said yacht which are now coming from Chatham in the Lighter hoy.' And on 23 August, 'his Majesty and the Duke of York came from Windsor and went to St James's to see her royal Highness and afterwards went down the river to Greenwich, where they saw a new yacht launched, and then in the evening returned for Windsor.'[22] The standing rigging was to be prepared in advance; the master attendant at Deptford was to be at Greenwich some days before the launch, 'in order to the landing of her shrouds fixed to the head of the mast, to be in readiness to set up as soon as she is launched.'[23] By the 30th it was reported: 'Her rigging is up, her sails are bent and her anchors and cables are all aboard.' The greatest delay, as usual, was 'the want of ballast and a commander and men'. The only other problem was the boat built by Pett, which had to be replaced by one made in Deptford yard.[24] However, on 11 September 'his Majesty and his royal highness went down the river to see the new yacht sail that was lately launched at Greenwich, and returned to Whitehall again the 15th.'[25] Her name, the *Fubbs*, was perhaps Charles's greatest attempt at control. It was the pet name of his mistress, Louise de Kéroualle, Duchess of Portsmouth, who already had one yacht named after her, and it might be seen as flaunting the King's masculinity. Yet the name and the ship survived through several regimes, being rebuilt under William of Orange in 1701 and again by George I in 1724, before being broken up in 1781. King Charles told Pepys that the *Fubbs* 'not only sails beyond anything he ever saw, but he believes to the highest degree that any vessel is capable of being brought' – for he was coming to believe that naval architecture was reaching the limits of possibility.[26]

Soon Pett was building another yacht on the site at Greenwich. The second *Isabella* was to be built for the Duke of Grafton, Charles's son by Barbara Villiers. It was slightly smaller than the *Fubbs*, 60ft long and 18ft 11in broad, so was narrower in proportion to its length. It was measured at 114 tons instead of 148. According to the shipwright William Sutherland, 'the *Fubbs* and *Isabella* yacht was at the building not only counted beautiful, but their lower bodies was admired, being … a perfect hanging conoid, with no other tails but what was purely requisite for the use, and the conjugates nearly circular; and the *Isabella* was well known to be an excellent sea boat.'[27] These were the last yachts of Charles's reign, and his brother would build no more.

Progress

Charles had seen the yacht anglicised with a deeper hull and more rounded features, though there was still an occasional tendency to revert to Dutch methods with the *Isabella Bezan*. The external decorations were now English in style, though the cabins might be decorated by paintings by the Dutch artists the Van de Veldes. The yachts had grown in size from the original *Mary* at 100 tons to her successor of 1677 at 166 tons. Two-masted ketch rig had been introduced. They were no longer regarded as river craft but were mainly used for passages along the English coast and across the English Channel and North Sea. But they retained the essence of the original concept, a highly decorated vessel with good sailing qualities, a large cabin for the principal passenger, and spartan conditions for the others.

5
The Loss of the *Gloucester*

Excluded from office in England by the Test Act of 1673 which banned Roman Catholics from holding office, James, Duke of York, was sent to Scotland, where his rule was condemned by his many enemies as autocratic and cruel. In 1682 Charles began to rehabilitate his brother. After two months in London, James was sent north again to wind up his affairs. On 3 May he left Windsor by coach and boarded a barge at Putney on the Thames west of London, 'attended … with all the barges of all the persons of quality about town.' They reached Erith – nearly twenty miles east of the city – by eleven and boarded the *Mary* yacht commanded by Captain Gunman, which was 'very plentifully furnished with all manner of necessary provision for him and his attendance tables for the voyage.' The ship had five hundred visitors according to Gunman, and he had to ask them to leave, 'for fear their weight should have forced the deck in.' At one o'clock they sailed for Margate at the mouth of the estuary.

Meanwhile, a squadron was being prepared at Portsmouth to meet them. The *Gloucester* was to carry the duke on the main part of the voyage and was fitted out, though Commissioner Sir Richard Beach complained constantly about the lack of money, furniture and kersey for equipping her. Nevertheless, 'the Duke and all the gentry went on board of her' and in two hours' time the baggage was out of the yachts on board the frigates. She was under the command of Sir John Berry.

The pilot on board the *Gloucester* for the voyage was a Scotsman, John Ayres, for 'his Royal Highness, having a particular knowledge of his ability and having done him many signal favours, trusted him'. Rather than passing outside all the sands along the route, he chose to follow the track used by colliers passing from the northeast to and from London.

At eight in the evening of Friday 5th, the fleet was off Lowestoft when the duke called Gunman under his stern, along with Sanderson of the *Charlotte* yacht. He asked them whether they could weather the Newark sand off Yarmouth without tacking. Gunman replied, 'no and thought the pilot was mad if he did not tack off', and Sanderson agreed. Ayres, however:

> … said we *could* weather the Newark and all other sands, and was much dissatisfied that any one should mistrust his judgement. His Royal Highness was pleased to answer, it would be a very secure way to tack and stand off *till twelve o'clock*, which the pilot very unwillingly agreed to. *At half past nine* the pilot very urgently desired to tack again, and his royal highness was still of opinion to stand off longer. The pilot answered, he would engage his life, that if we tacked we should weather *all* the sands. Notwithstanding his arguments his highness commanded the pilot to stand off a glass [half an hour] longer: at ten we tacked and stood close hauled N by E. All night we steered NNE till two next morning; then we steered N and at four NNW the pilot constantly affirming that this course would carry the ship out of all danger.[1]

Mary was sailing at least half a mile ahead of the flagship on her lee bow, and Gunman denied that he had any instructions to guide the other ships. At four in the morning he was confident enough to leave the deck in the charge of his master's mate, William Sturgeon, with orders to 'observe and follow' the movements of the *Gloucester*, and 'if he come in shoal and unexpected depth he should immediately come and call me, and bear up round'. Seaman John Eaton was on the tiller, and William Shepherd was in the chains to one side of the ship, casting the lead to report on depths of water. He made three casts of the deep sea lead, then used the hand lead as the water beamed shallower, though not dangerously so. All seemed quiet; Samuel Pepys was in bed on board the *Katherine* and wrote later:

> Little sign of our commander thinking the *Gloucester* in any great danger, when not only the Duke and Colonel Legge but Sir John Berry himself were gone to bed … and the pilot gone to his cabin, if not in bed … Captain Gunman also and Captain Sanders [of the *Charlotte* yacht] … were both in bed. Whether Captain Birch and Captain Leake [of the – and –] were up or no I know not …[2]

However, they were approaching a particularly dangerous sand. According to John Seller:

> The Orrey is a sand that beareth east by north, half northerly, from the Haseborough Church about seven or eight leagues off, and is about two miles from the Lemon: between which two sands there is a channel, where you will have about 17, 18 and 19 fathom water.
>
> … but myself was a-bed and was wakened by Captain Leake's crying out that the Duke was on ground. So that Captain Wyborne of the *Happy Return*, who was about a mile astern of us, who was the only commander of the whole fleet that was out of his bed at that time.[3]

At 5.30am Captain Gunman was called on deck, to be told that the depth was now seven fathoms, or 42ft, and they were clearly approaching a bank. He kept the lead sounding and tried to find out if they were to windward or leeward of ten shoal, finding it was the latter in two or three minutes. He hoisted the standard daytime warning signal, a jack above the poop, but he did not fire a gun immediately as the armament was apparently not ready and the gunner was not on deck. But by this time the *Gloucester* was aground and fired a gun herself.

It did not seem particularly dangerous at first, according to Berry:

The ship beat long the sand, not sitting fast; while our rudder held we bore away west, and upon every lift of the sea went off; at last a terrible blow struck the rudder, and, as was believed, struck out a plank near the post, as the ship made eight feet of water in an instant.

Berry tried to persuade the duke to take to his boat, 'but his highness was unwilling to consent to, hoping, as Sir John did, the ship might be saved'. But water continued to enter and:

no manner of hope being left, Sir John did again, with all manner of earnestness, request his royal highness to go off in his boats and the yachts; to which his royal highness consenting, the barge was hoisted out, and his royal highness took as many persons of quality with him in the boat as she could carry ...[4]

A rather misleading view of the wreck of the *Gloucester*, with James and his entourage escaping by boat. In fact the Duke of York probably embarked straight into the boat from the *Gloucester*. (© National Maritime Museum, Greenwich, London, BHC3369)

THE LOSS OF THE *GLOUCESTER*

That proved to be a very controversial point. John Churchill, later the Duke of Marlborough, was a passenger in the *Gloucester*. He was saved and related the story to his wife soon afterwards:

> ... all those he mentioned that were drowned were lost by the Duke's obstinacy in not coming away sooner; and that was occasioned by a false courage to make it appear, as though he had what he had not; by which he was the occasion of losing so many lives ... But when his own [life] was in danger, and there was no hope of saving any but those that were with him, he gave the [future] Duke of Marlborough his sword to hinder numbers of people that to save their own lives would have jumped into the boat, notwithstanding his Royal Highness was there, that would have sunk it. That done, the Duke [of York] went off safe.[5]

This has to be treated with a certain amount of caution, as it was written down by the Duchess many years later, and it was not necessarily unbiased – Marlborough would later desert James during the Glorious Revolution of 1688 and had plenty of motive for denigrating his character. Bishop Burnett, in his *History of His Own Time*, went as far as to say, 'The Duke got into a boat: and took care of his dogs, and some unknown persons who were taken from his earnest care of his to be his priests: the long boat went off with very few in her, though she might have carried above eight more than she did.'[6] This was hotly disputed and the whole affair tended to damage the duke's reputation, which was already very controversial. He had none of his brother's charm and his Roman Catholicism was deplored by the great majority of the population. His biggest asset was his integrity, but that was challenged in some accounts of the affair.

In any case, the *Gloucester* was definitely sinking. Berry escaped down a rope over the stern into Captain Wyborne's boat, but the Earl of Roxburghe, Lord O'Brien, Lord Hopton, Sir Joseph Douglas and the king's brother-in-law, Mr Hyde, were the 'persons of quality' lost because they could not be found in the confusion, along with about 130 seamen.

In the meantime, James was rowed over to the *Mary*, which had turned back towards the sinking ship. He and his retinue were brought on board and Gunman set sail to the north, followed by the *Charlotte* and *Katherine*. At eight in the evening of Sunday the 7th, 'I came to anchor in Leith Road where I safely landed his Royal Highness, God be praised.'

The *Mary* stayed on at Leith while James completed his business over the next week. James had clearly not lost confidence in Gunman, as he transferred him temporarily to the command of the optimistically named *Happy Return* for the voyage back, with Captain Lawrence Wright in charge of the *Mary*. They had a slow and presumably cautious voyage and arrived off Tilbury in the Thames on the 26th. Again Gunman praised God and resumed command of the *Mary*. But his troubles were just beginning.

In the toxic political atmosphere of the early 1680s, scapegoats were needed for the loss of life on the *Gloucester*. According to E Ridley, 'the pilot is a known republican, ... it's not only suspicious but evident he designed his ruin with the whole ship, having made a provision for his own escape, but he is taken and will be tried for his life.'[7] It is far-fetched to believe that a professional pilot would deliberately wreck a ship for political reasons, and assume that he would escape while the duke would not, and in any case the sinking was far from a foregone conclusion, even after the ship went aground. Nevertheless, Ayres was put on trial and harshly sentenced to life imprisonment, though he cannot be said to have been in full control of affairs during the voyage.

If the reports of Ayres's politics are correct, then Gunman was the direct opposite. Far from being a republican, he was a long-standing protégé of the Duke of York. Born in 1634, he was a captain by 1666. He first took command of the duke's own yacht, the *Anne*, in 1669 and consistently referred to it as 'His Royal Highnesses' yacht'. He was appointed second captain of James's flagship, the 100-gun *Prince*, during the Third Dutch War, in effect the commander of the ship, as the first captain was the duke's chief of staff. He took command of the *Anne*'s successor, the second *Mary*, in 1677 and maintained his link with the duke. It was probably an attempt to undermine the duke that brought Gunman before a court martial.

The luckless mate William Sturgeon had already been arrested and held in the Marshalsea prison, where Gunman believed he had been put under great pressure. Gunman heard of 'the said mate's wife running up and down to procure people to swear her husband against me and prove such as she brought to her husband against me', though without much success. A court martial convened on board the yacht *Charlotte* on 13 June with the bluff tarpaulin sailor Sir Richard Haddock presiding, for 'enquiry into the loss of his Majesty's ship the *Gloucester* Sir John Berry Knight commander on the Lemon Sands the 6th May last'. Gunman, who had received no indictment and had not prepared a defence, soon found that the tide was running against him. It was implied that the *Mary* had been directly ahead of the *Gloucester* at the time of the accident, rather than off the bow. He soon found himself accused of neglect of duty in failing to fire a gun as a warning. He pointed out that he was not a pilot, and that 'I never was a pilot, nor ever received pilot's wages, or got five pence by pilotage.' Sturgeon the mate, Shepherd the leadsman, Eaton the helmsman, and seaman Edward Reynolds gave evidence, but they said little of consequence. Nevertheless, the court acquitted Sturgeon, but concluded that Gunman was guilty of neglect of duty 'in meeting with shoal water and not firing a

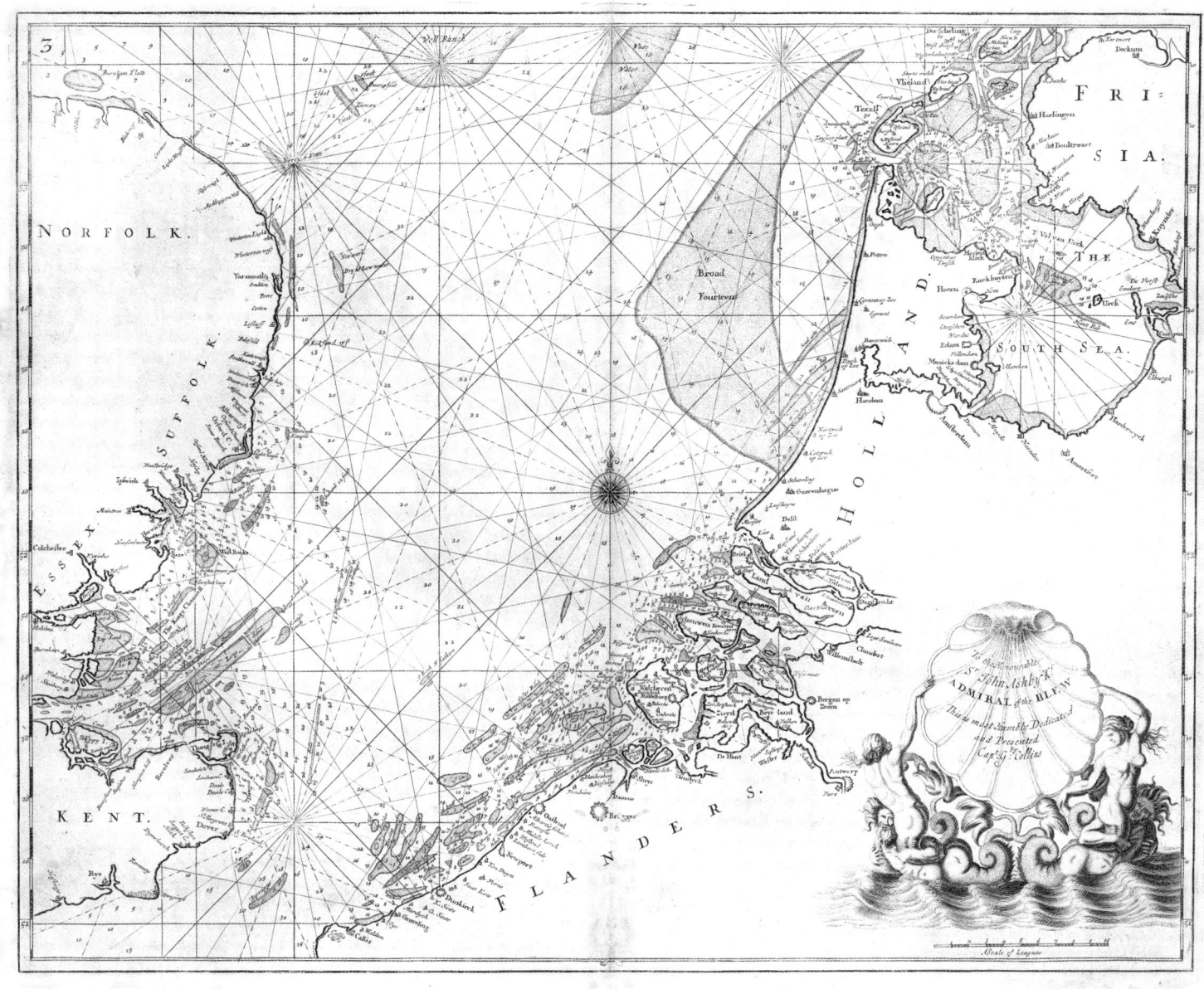

Greenville Collins's chart of the North Sea, showing the sandbanks off the Thames Estuary and the Lemon and Ore banks towards the top right.

gun (which is the usual signal given in cases of danger and shoal water)'. He was sentenced to be dismissed the ship, to be imprisoned 'during his Majesty's pleasure', and to forfeit a year's pay to the Chatham Chest, a charity for seamen.

A shocked Gunman was taken off to the Marshalsea Prison to begin his sentence. At Windsor, the King knew of this by the end of the day and reacted quickly. He wrote to the Admiralty asking for 'a particular account of all the proceedings and proofs before the court-martial relating to him, that his Majesty may understand the true state of his case' and directed that 'such imprisonment shall be in his own house at Deptford', and that the command of the *Mary* should not be filled until the matter was resolved. That did not last long – on the 19th the Admiralty Board was told that 'His Majesty ... extends his pardon to Captain Gunman, the forfeiture of a year's pay ... only excepted, his meaning being that he be forthwith released of his imprisonment and restored to his command of the *Mary* yacht.'[8]

He was back in command of the *Mary* by 23 June and soon escorting the King to Chatham in the *Charlotte*, with other yachts. Samuel Pepys continued to blame him. He resented not being called as a witness at the court martial, and wrote of 'the evil occasioned by Captain Gunman and the pilot's misbehaviour on that occasion.'[9] But Gunman continued to ferry diplomats and courtiers until he died in his house in Deptford in March 1685, a few weeks after his patron succeeded to the throne.[10]

~6~
The Yachts at War

~ The Second Dutch War ~

In 1664 Charles was forced into a war with the Dutch, whose navy had already been defeated by Cromwell's ten years earlier. The yachts were allocated humble tasks during this, the second Anglo-Dutch War. At the end of June 1666 the *Henrietta* and *Monmouth* were both ordered to deliver soldiers to serve as marines in the fleet. A few days later the *Henrietta* was to deliver eighty men, probably pressed, to Rear Admiral Sir Joseph Jordan.[1] In September the *Katherine* was to accompany the *Elizabeth* hoy to Rye, to take 'the men they can conveniently carry' out of two ships and bring them to the main fleet. The *Monmouth* also had other duties, landing Sir Thomas Clifford at Dover in September, and then sailing off Dungeness for two days to 'acquaint all such ships, frigates and other vessels, that come with intelligence, that they repair to the fleet plying on the French coast.'[2] Occasionally they saw action, as in April 1665. 'The *Drake* having taken a small steigerschuit of the Brielle that was a-fishing and the afternoon the *York* [*Anne*] pleasure boat brought another.'[3]

The Dutch wars were mostly fought in the southern North Sea within easy range of London and the Thames, and between actions the fleets anchored at the Nore, the Downs and the Gunfleet off the Essex coast. This was convenient for the King to visit. Thus on 3 May 1666: 'His Majesty came aboard the *Charles* with his Royal Highness and the Duke of Monmouth, Earl of Bath, Lord Berkeley and many followers.' They held a council of war next day, and:

> After that his Majesty and Royal Highness went aboard the *Monmouth* pleasure-boat to go to meet the *Defiance* coming down wind westerly. After that they sailed to the eastward to view the fleet, where the sailors showed their joy by great shouting and throwing away their caps. His Majesty was at Sheerness.

Again on the 5th:

> … in the morning by 4 o'clock. The pleasure-boats went to the eastward to meet the *Rupert* and were much satisfied with her, then came aboard the *Royal James*, where they ate some cake and drank a glass of Canary, thence to the *Prince* and so to dinner. Afterwards they went a-trawling and got some fish. The wind easterly, they went aboard Sir Edward Spragge [in the *Triumph*]. They came aboard the *Charles*, stayed half an hour, then aboard their pleasure-boat, where we kissed his hand, and so took leave. This was near 7 at night.[4]

The war ended in 1667 after the Dutch raid on the Medway ended English hopes of a decisive victory.

~ Pretext for war ~

In August 1671 the *Merlin*, commanded by Captain Thomas Crow, was sent to bring the ambassador's wife, Lady Temple, back from Holland. On the 14th she passed through a large Dutch fleet of thirty-seven ships under the great Admiral De Ruyter in the middle of the North Sea. Crow probably thought himself bound by orders issued by the Duke of York in 1664

'concerning the exacting from the Dutch fleet by a single ship (and that perhaps a small one) the striking of their flag and topsails'. It was best to avoid such a situation:

> but if it happen that such a case fall out, I hold it necessary that the commander of the single ship, how small soever, do exact it from them by the exchange of some shot, though I do not hold it a duty for him to be sunk upon such unequal terms …[5]

Crow fired 'several shotted guns' which De Ruyter was not in a position to reply to immediately. His second in command Van Ghent took it for a salute and fired seven unshotted guns, followed by De Ruyter with nine. Van Ghent received a demand to strike his flag and lower his topsails in formal salute and went to the *Merlin* to inquire what 'person of quality' was on board to justify the flying the Union flag at the masthead. He knew Lady Temple, but pointed out that his orders did not allow him to accede to such a demand. Crow felt himself obliged to accept this, but there was outcry in England. He was put in the Tower of London and later released but not employed again. Charles was already resolved on war with the Dutch after the secret Treaty of Dover with Louis in 1670, and this seemed a possible pretext. However, the Dutch were conciliatory.[6]

A royal visit to the fleet, probably in June 1672. The *Cleveland* and *Katherine* are seen to the left. The Admiralty flag, Royal Standard and Union flag are hoisted on the masts of a three-decker, probably the *Prince*, the Duke of York's flagship. (© National Maritime Museum, Greenwich, London, BHC0299)

THE YACHTS AT WAR

The Third Dutch War

Instead, Charles launched an unprovoked and largely unsuccessful attack on a Dutch convoy returning from Smyrna, this time leading to war in 1672. Charles visited his fleet more often during the Third Anglo-Dutch War. Perhaps it was because he had played a large part in instigating this one, rather than having it forced on him. Perhaps he also felt it was necessary for him to conduct relations with the French allies.

The Duke of York was in personal command as Lord High Admiral in the early stages of the war, and on 2 April 1672 he used his own yacht to join the fleet in the Medway, despite the bad weather. 'This afternoon it blew hard in gusts. This evening HRH the Duke of York came into this river in the *Anne* yacht and anchored near the *Prince*. … His Highness lay on board the *Prince* tonight and the nobility that came with him'.[7] The King visited three weeks later on Tuesday 23rd. The Duke went out to meet him, and an elaborate array of flags was hoisted when he came on board the flagship *Prince*. That night, 'the King supped on board the *Prince*, and then went on board the *Cleveland* yacht and lay there all night. Fair weather tonight'. He often slept on board his yachts in fair weather and did so for the next two nights, which gives some idea of the standard of comfort; but on the 26th, after the fleet moved out into the Thames Estuary: 'This afternoon and tonight it blew a strong gale at W. This night the King lay aboard the *Prince* and the Duke also.' Finally, on the 27th: 'This afternoon the King took his leave of his Royal Highness and went aboard his yacht and turned up the river of Thames, the wind at [west by north], a fine gale'.[8]

There was a short visit on 5/6 June, and on the 18th the fleet was at the Nore when:

> At 7 o'clock the King and Queen came aboard the *Prince* and several of the nobility. His Royal Highness received them with a bountiful entertainment, having all the colours and pendants flying … This evening the King and Queen and his Royal Highness supped on board the *Victory* … The wind blew hard westerly that night.

Perhaps that is why the Queen slept in the *Prince*, while the royal brothers retired to their respective yachts. And on the 19th: 'This morning the King and Queen and all the nobility went from aboard the *Prince* and visited several ships in the fleet, and from thence went to Sheerness, and so up to London that night.' But the King was back less than a week later, on the 25th. 'This morning his Majesty came from London; this evening his Majesty came down in his yacht … the King supped on board the *Prince* and lodged on board his yacht. Fair weather tonight. We rode fast.'[9]

There was another visit on 24–26 August when the fleet was off Harwich. The King was back again on 9 September when the ships were back in the Thames Estuary after a largely unsuccessful summer campaign. There was the usual display of flags.

> This afternoon at 4 o'clock the King came aboard, and Prince Rupert and several of the nobility were with His Majesty. When his Majesty came within two miles of the *Prince*, his Royal Highness commanded the standard to be struck, until such time as his Majesty came aboard. At the striking of his Royal Highness's standard all the flags in the fleet were struck immediately, and kept down until the standard aboard the *Prince* was hoisted; then they hoisted theirs.

The brothers indulged in a little hands-on hydrography, 'This morning his Majesty and [his] Royal Highness went in their yacht to the Red Sand and took a view of that channel, and came aboard again before dinner time.' The King slept aboard the yacht as usual with his nobles, but Admiral Sir John Narborough was concerned about security. 'I ordered my lieutenant to guard the yacht with a pinnace and 20 men with small arms, not to suffer any boats to come near the yacht where the King lieth.' This was repeated until he returned to London on the 13th.[10]

In 1673 the Duke of York was forced to give up command when the Test Act made it impossible for Roman Catholics to hold office. The King was on board the fleet on 1 May at the Nore when Sir Edward Spragge arrived to take up his command, but he soon left.[11] By the middle of the month the fleet was anchored in Rye Bay on the south coast to link up with the French under Vice Admiral Comte d'Estrées, who arrived in the morning of the 16th. That evening the King and duke came out from Rye in their yachts, including the *Cleveland*. On the 17th they held a council of war on shore and on the 18th, 'His Majesty and Royal Highness still in the fleet; was by six under sail in his yachts turning to his French auxiliaries, where he dined on board the French Vice Admiral le Comte D'Estrées. He went the same day to Rye.'[12] There was another brief visit on 11/12 June, this time at the Nore.[13]

Another voyage began on 6 July, illustrated in great detail by the artist Willem Van de Velde the Elder, who had recently been lured from his native Holland. Based in the *Kitchen* yacht while the King sailed in the *Cleveland*, he made numerous sketches of every stage of the trip. The King boarded the *Cleveland* at Woolwich and sailed in light westerly winds accompanied by the *Kitchen* and another yacht. They were saluted by various merchant ships during the passage and were cheered by soldiers in Tilbury Fort before anchoring briefly in Gravesend Reach. Food was served from the *Kitchen* in Sea Reach while they were under way. Prince Rupert joined at

The Battle of the Texel in August 1673 by Van de Velde. The small vessel in the very centre of the picture is possibly the *Henrietta* sinking. (© National Maritime Museum, Greenwich, London, PAJ2530)

Sheerness and a French officer came on board. A great number of boats came alongside and the King was rowed ashore. Next morning they began to tack down the estuary and the King came out by boat for a council of war on board the *Royal Sovereign* about how to attack the Dutch in their secure anchorage at Schoonveld.[14] 'The King and his Royal Highness etc, returned to London about 9 o'clock, being the time of the tide.'[15]

The King boarded the *Cleveland* again at Woolwich a week later, accompanied by the Duke of York. The yachts were becalmed for a time as recorded by Van de Velde, who made ninety drawings. A light breeze got up and at midday they were fed, while still under way, from the *Kitchen*. The combined fleet was in sight from Sea Reach, where the yachts anchored. The King went on board a snow, a two-masted squarer-rigged vessel, and went on board the *Royal Sovereign* again.[16] A council of war decided not to attack the Dutch in the Schoonveld 'till further order from the King upon no pretence whatever.'[17]

At the Battle of the Texel on 12 August 1673, the *Henrietta* yacht was the only ship sunk on either side and none were captured. Details of the loss are sparse but it was implied that it was partly due to the captain staying close to Admiral Ossory, and a poet mocked Dutch triumphalism.

> Yet e're he parts the *Henrietta* yatch
> The last effects was of his fury taught;
> This little vessel full of spirit, gay
> Without, as any lady of the May;
> Within deckt a valiant soul whose heart
> Could not admit from his admiral part;
> His love adventring past his strength, is drench,
> And in the briny wave his passion quench.
> Poor triumph! where odds so mighty were;
> Yet Amsterdam shall mak't a man of war.[18]

The battle was a tactical draw, but strategically it was a victory for the Dutch, and there were divisions between the English and their French allies. The war was unpopular in England when a revived France seemed far more dangerous, and any hopes of an easy victory had been dashed. Parliament demanded peace, which was concluded early in 1674, leaving the French to fight on.

James II

James II of England and VII of Scotland came to the throne on 6 February 1685 on the death of his brother. As Duke of York he had served as Lord High Admiral and taken part in many voyages in the yachts, but he saw far less of them during his reign. It started with rebellions by the Earl of Argyll in western Scotland, and his nephew, the Duke of Monmouth, in southwest England. The yachts were deployed during the latter affair. Captain John Walford of the *Merlin* was ordered:

> to ply to and fro between the Needles and Hampton River to try and meet with and search all such vessels you find coming out of the Hampton River, Lymington or any other creek if you suspect them to have any rebels or disaffected persons to the King and government on board them. You are to seize all such persons and vessels making their escape and bring them to me. If they resist you are to take, burn or destroy all such vessels or person that are trying to escape out of the kingdom at any other part or place.[19]

James, Duke of York, holding the baton of the Lord High Admiral, by Pieter Cornelisz van Slingeland. (Royal Collection Trust / © Her Majesty Queen Elizabeth II 2022, 422088)

The *Cleveland* was ordered to take the Earl of Bath from Portsmouth to Exmouth, but Captain Wright was 'unacquainted with going out at the Needles' and had to employ a pilot. Both rebellions were soon crushed, but the punishments inflicted by Judge Jefferies at the Bloody Assizes increased the unpopularity of the King.

James's main excursion round the kingdom visited several ports, but it was done by land. He was focused on establishing religious toleration, which included many appointments for Roman Catholics, and on building up an army, much to the disgust of his most powerful subjects, who remembered the tyranny of Oliver Cromwell, and feared the establishment of a repressive Catholic regime like that of Louis XIV across the Channel.

The Glorious Revolution

The crisis of the regime came in 1688, after James had alienated the Church of England and most of the gentry. His son-in-law, William of Orange, threatened an invasion and a fleet was fitted out to oppose it. It included three yachts, the *Cleveland*, *Katherine* and *Kitchen*, among a total of sixty-one ships ready or fitting out. This perhaps indicates the shortage of small vessels for message-carrying and reconnaissance. Besides the yachts, there were only three sixth rates which were suitable for reconnaissance duties, including the *Saudadoes*, which had been reclassified.[20] Unlike the experience of the three Dutch wars, it was not easy to find the opposing fleet, as William tried to evade James's forces to find a site for landing.

James's fleet under Admiral Lord Dartmouth anchored in the Gunfleet to the north of the Thames Estuary, and on the 20th the *Cleveland* was sent into Harwich with letters. On the 30th Dartmouth sent the *Katherine* to the east, the *Kitchen* to the south and the *Saudadoes* to the north. On 4 November the

The arrival of the yacht *Mary* at Gravesend in February 1689, carrying Queen Mary to complete the Glorious Revolution. The white standard bears the inscription, 'For the Protistant religion and the libirty of Enclant.' (Royal Collection Trust / © Her Majesty Queen Elizabeth II 2022, 912851)

Katherine returned 'with an account of a considerable fleet being passed to the westward on Saturday.'[21] William had succeeded in evading the English fleet, and was heading down the English Channel some days ahead of Dartmouth. William had already landed his forces at Torbay by 16 November when the *Cleveland* was sent into Portsmouth asking Commissioner Sir Richard Beach to send out what ships he could to strengthen the fleet.[22]

Large elements of James's army, including his son-in-law, George of Denmark, and Churchill, defected to William. As the regime began to collapse, the new rulers were worried about important personages escaping to France where they could set up a government in exile. The *Fubbs* was stationed at Portsmouth and was intended to ferry Lady Scott and her family to Guernsey. Commissioner Beach was informed that 'there is a titular [Roman Catholic] and some priests that embark themselves in the *Fubbs* with Lady Scott. I have therefore desired Sir William Jennings to search for them before she sails'.[23] There was particular concern about the King's infant son, whom the opposition believed was spurious and had been smuggled into the Queen's bedchamber in a warming pan. The *Mary*, also stationed at Portsmouth, attracted suspicion. Captain Cornwall wrote that:

> going ashore to get provisions he saw goods going on board the *Mary* yacht. On enquiring he was told they were Lord Dover's and that the Prince of Wales was going; and he saw several women going on board who were said to belong to him … Sent his coxswain to enquire, and they told him they were bound for St Malo as soon as the moon was up; but the milk and pap were sent for ashore and he believed they would not go that night.[24]

Guards were posted to prevent the prince from boarding.

Meanwhile, King James attempted to flee the kingdom in an echo of his brother's escape of 1651. His small vessel anchored off Faversham in the Thames Estuary on the night of 11 December and was boarded by local fishermen who handled him roughly – an event he never forgot nor forgave. William called him to London on the 16th, but he was allowed to flee again on the 23rd, avoiding any repetition of the trial and execution of Charles I.

In February 1689 Admiral Herbert was sent to Holland with a fleet, including several yachts, to fetch William's wife Mary, who as the eldest daughter of James had a much better claim to the throne than William himself. In the Thames Estuary she transferred to the yacht *Mary*, which had been named after her, and she was depicted passing Gravesend on 11 February with the wind astern and flying a flag bearing the slogan, 'For the Protestant Religion and the Liberty of England.'[25] She landed at Greenwich on the afternoon of the 12th 'and was 'received with very great joy, demonstrated by the ringing of bells, bonfires, etc, and was waited on by most persons of quality in town'.[26] The joint monarchy of William and Mary was proclaimed by Parliament the next day and the Glorious Revolution of 1688 was complete; under the settlement, in future monarchs would have to work hand in hand with Parliament. It also began a long era in which France was the principal enemy, for war was declared in July.

Part II

To the Wars

1
The Changing Role

The function of the royal yachts changed as soon as King James was deposed in 1688. For one thing, no sovereign until the nineteenth century approached Charles's and James's love of the sea, and the yachts gradually became more functional, though not without elaborate decoration in many cases. But from 1689 until 1837, every British sovereign except Queen Anne had a European territory to govern and that changed their focus. Wars became more intense, wider ranging and longer – about two-thirds of the reign of William of Orange was in wartime, nearly all that of Anne, a third of that of George II and more than half of that of George III – only George I had a relatively peaceful reign. Royal participation in the wars varied over the years: William of Orange had to travel to his campaigns in Ireland and Flanders, and much later George II went to Germany to lead his army at Dettingen in 1744, but on the whole it tended to decline.

King William

At the beginning of the war in 1689 there was a move to turn some of the yachts into minor warships, though only the *Portsmouth* had a full-scale conversion to a bomb vessel in 1694, which must have involved a good deal of new structure. Other yachts did not stay out of action. The *Charlotte* was charged with maintaining communication with Guernsey in 1689, which involved carrying high officials at one level, and soldiers' clothing and prisoners on another. But in May 1689, soon after war with France was formally declared, she was searching suspected French ships off Beachy Head. Later in the month, despite transporting Sir Edward Carteret as a passenger, she chased a small French privateer, firing at and capturing her.[1]

Though he was reluctant to engage in a campaign in Ireland, William could not ignore the presence of a Jacobite army led by James in person. He travelled to Chester and on 11 June he embarked on the *Mary* yacht under Greenville Collins to a salute of thirty-one guns – his brother-in-law, Prince George of Denmark, was already on board the *Henrietta*. There were five more yachts escorted by six warships under Sir Cloudesley Shovell. Progress was slow in light winds and ships were separated in fog, causing some anxiety, but on the 14th, 'the wind grew high and pushed forward and so that at about half an hour after one, the King cast anchor.' The King was rowed ashore in Shovell's barge to another 31-gun salute and landed at Carrickfergus.[2] William won the Battle of the Boyne on 1 July 1690. James retreated and William's forces had the upper hand, so he was able to leave Ireland late in August. However, some of the yachts were still deployed to maintain contact between the two islands.

To the war in Flanders

King William was keen to play his part in the continental war against France, which was largely being prosecuted from his homeland in the Netherlands. By 28 November 1690 he was 'daily expected' there, but the winter was far advanced and it was early in January before he left London. Admiral Sir George Rooke was chosen to head his escort as 'no body was thought more proper to be trusted with the guard of his royal person in that voyage'.[3] The son of a gentleman of Kent, Rooke was a captain by the age of twenty-three, commanded the 54-gun *Deptford* during the revolution of 1688, and escorted Queen Mary from Holland shortly afterwards. He escaped censure for his conduct at the defeat at Beachy Head in 1690. He was 'a gentleman of very good parts. Speaks little, but to the purpose … a stern-looked man of brown complexion, well shaped'.[4]

It would be the first time that a monarch had made such a journey in the yachts in wartime and it was not easy. The King left Lambeth by coach on 6 January and spent the night at Sittingbourne, intending to embark at Margate. The wind, however, was contrary and he went to Chatham to view the ships there, then to Canterbury, where he heard that the wind was set easterly and a hard frost was coming in; he was back in London by the 9th awaiting a change of weather. He left again on the 16th and at Gravesend he embarked in the *Mary* yacht; next day they were clear of Margate and heading for Holland. The Earl of Nottingham, Secretary of State, was detained in

London to give evidence at trials for treason but set off in the *Saudadoes* on the 21st.

The King arrived off the Dutch coast on the 19th in poor visibility. A fisherman informed them that they were close to the coast and the King, impatient after so much delay, got into a shallop to be rowed ashore, with some of his noblemen in another. The fog thickened, they lost sight of the ships and it was soon clear that they were much further off than they thought. They were among ice floes and decided to lie still for the night. They reached Goree next morning and the King went ashore, then by boat again to meet the deputies of Holland.[5] When the Queen heard of it she ordered thanksgiving in the churches: 'We also praise thy Holy Name, for that wonderful preservation which thou hast lately given him, at sea, in saving him from so many deaths, as were ready at once to have taken him from us, who is the breath of our nostrils, and to have quenched the light of our Israel'.[6]

He could not stay long – on 24 February it was reported that he was hunting near his palace at Het Loo, but intended to sail for England in a few days. The government in London was preparing '[t]he list of the ships that are to be set out in Holland', while on 6 March the King conveyed 'his pleasure … that the same number of men-of-war and yachts as the King had to attend him to Holland … be forthwith made ready in order to be on the coast of Holland this day week to attend him to England.' The campaigning season was approaching and there was a shortage of ships, but Rooke 'whom we have appointed to attend his Majesty on his return' was ordered:

> to repair to the Downs and hoist his flag on board the *Lion* or *Charles* galley, and sail with the ships and yachts mentioned in the margin to the Brill in Holland, and, when arrived, to order the yachts to Rotterdam and the *Saudadoes* to Flarden and the rest of the ships to Goree. Upon his arrival in Holland he is to give you notice thereof and follow the King's orders.

Rooke sailed on 17 March with four men-of-war and two yachts. This time the voyage was uneventful, the King dined on board the *Duke* at the buoy of the Nore, and was back in London by the evening of 13 April.

Back to Flanders in May 1691

He did not stay long there either. On 24 April he 'declared in Council his resolution of going to Holland on Friday next.' Rooke took command again, the King left Kensington on 1 May and arrived at Harwich at six that evening. They set sail next morning with a fair wind and arrived in the Maas at six on the 3rd after a fast passage. This time he spent a full season in the theatre of war and in October 1691 William began to prepare for his return. He intended to travel in the *Greyhound*, which Charles had used in the past and ordered it to be prepared 'for the transporting to England the persons, servants and goods belonging to me and my office'. Orders to accompany him were issued to four other vessels and the captains of the yachts *Henrietta*, *Katherine* and *Fubbs*.[7] The flotilla found a favourable wind and arrived off Margate after about twenty-four hours, confusing the courtiers, who went to Harwich expecting to meet him. This time the danger came by land: at Gravesend his coach overturned but he was not hurt.[8]

By 1692 the voyages were beginning to settle into a kind of pattern. Timing was constrained because the King had to be available while Parliament was sitting in the winter and spring, but that fitted in with the need to be present in Flanders during the summer and autumn campaigning season. On 27 March 1697, for example, it was reported that the King was 'designing to go to Holland as soon as parliament is up'. He left in March, April or May for the start of campaigning, and came back in October or November.

The Admiralty was usually given adequate notice of the King's intended movement, which was often increased by delays for one of two reasons. The King tended to wait for a suitable wind and there was even a primitive form of weather forecasting. In October 1694 he was still at The Hague with much rain, and it was hoped the wind would turn 'for which his Majesty waits'. The theory of depressions was still unknown, but sailors had observed that there were westerly winds as the rainy warm and cold fronts passed over, and then the wind would veer to the north. Since the prevailing wind was from the southwest, delays were more likely and more prolonged on the trip from the Netherlands to England.

The King was not inclined to wait on board if the wind was unsuitable. At Gravesend in May 1694, 'he endeavoured to put to sea, but the wind being contrary, he is gone to Canterbury to wait till it turn fair for his going.' On Thursday, 30 April 1696 the King's baggage was embarked at Margate and it was intended that he should sail on Saturday. He duly left London at seven in the morning of the 2nd, 'to lie this night near Margate'. The yachts set sail next day but were forced back by contrary winds. Rather than go back to London, the King had the Lords Justices attend him at Margate as he awaited a wind.

Secondly, the King was often delayed due to urgent business on one side of the North Sea or the other. In October 1692 '[o]n the news of the French investing Charleroi … the King immediately returned to the army; and the forces embarking for England were countermanded; so not certain when the King will come'.[9] On 13 February 1694 it was reported that he planned to leave England at the latter end of the following month. On 29 March he declared he was going by the 14th of next month and ordered his equipage to be got ready. On 26 April he went to Parliament to settle business before going, and on the 30th he went from Kensington to Whitehall, then to Gravesend to embark, having signed numerous warrants before leaving.

Since the Queen was joint sovereign, it was only natural that she exercised nominal power during the King's absence, though she was reliant on a council with a strong naval and military content. In 1691 the nine members included George, Pembroke, the Earl (as he then was) of Marlborough and Admiral Russell. Mary died at the end of 1694 to William's great grief. Rather than appoint a single person as regent during his absences, he used the 'Lords Justices' which in 1695 comprised the Archbishop of Canterbury, the Lord Keeper of the Privy Seal, the dukes of Shrewsbury and Devonshire, the earls of Pembroke and Dorset, and Lord Godolphin. The Duke of Norfolk threatened to resign his offices at not being included, but retracted the threat.

William normally used the *Mary* until the purpose-built *William and Mary* was ready in 1694. Accommodation was cramped outside the royal quarters, so important civil and military leaders travelled in the other yachts, with the *Katherine*, *Fubbs* and *Henrietta* being used during this period. The yachts sailed with a substantial escort of regular warships, partly for protection against enemy raids and partly to add to the grandeur of the occasion. In February 1691 the French were supreme at sea after their victory at Beachy Head in the previous year, and the Admiralty pointed out that deploying 'the ships that are to attend the King in his return from Holland' would weaken the forces elsewhere. For the outward voyage in 1692 Shovell had ten men-of-war, and later in the year he sailed from the Downs with five to bring the King home. In October 1693 Admiral Mitchell had four third rates, four fourths and the *Charles* galley, besides several yachts. In April 1694 Lord Berkeley had twelve ships. On 12 April 1697 the Admiralty wrote to the Navy Board:

> Very urgently requiring the Yachts *William and Mary*, *Fubbs*, *Charlotte*, *Isabella* and *Navy* to be put in condition for sea and those that need cleaning to be cleaned in order to accompany the King on passage to Holland together with the *Henry Prize*, *Lizard*, *Royal Transport* and *Essex Prize* which are to be provided with provisions and Stores.[10]

Sir Cloudesley Shovell commanded on several occasions. He had risen from humble origins to command a ship and he played a key role in keeping the English fleet inactive during William's invasion of 1688; he was a staunch Whig who could be trusted, a good seaman and was noted for 'his courage, capacity and honesty which it will be hard to parallel in another commander.'[11] Admiral Sir David Mitchell, who commanded the escort in 1693 and conducted Tsar Peter during most of his trip to Britain, was another 'tarpaulin' officer who had risen through the ranks. He was said to have been apprenticed to a Scottish shipmaster and pressed into the navy. He rose to captain 'without any recommendation but his own merits', and went into exile in Holland during the reign of James II and learned Dutch – his loyalty to William's regime was not in doubt. He was 'a just, worthy man of good solid sense, but extremely afflicted with the spleen, which makes him troublesome to others as well as himself.'[12]

The King usually sailed from Margate or Harwich to shorten the sea crossing; the choice of departure point was still undecided in May 1695 and royal guards were sent to both places.[13] The yachts could be anchored at Margate Roads with the escort in the Downs, or in Harwich Harbour with the warships anchored outside in the Gunfleet or Rolling Grounds. They usually landed at Brielle on the mouth of the River Maas, which gave quick access to the Dutch capital at The Hague. It was not without its hazards. In October 1693 it was reported, 'the King's Dutch secretary going on board in the boat of the ship, the sea running very high, and thinking the boat was sinking, leapt overboard, and was drowned'.[14] Conversely, in May 1694 the King 'was rowed six leagues in the pinnace of a man of war, the wind being contrary, and came there about midnight.' Alternatively he might land at Oranje Polder in the Westerschelde, which was closer to the battlefields to the south, as he did in 1692 and 1697.

The Queen and the Lords Justices often awaited news of the King's return. In October 1692:

> An express came in this morning from the Bailiffs of Yarmouth with an account that a tender had come in there belonging to the *Kent* and informed them that his Majesty came out of the Maas [Meuse] on Saturday last about noon in the *Mary* yacht attended by two English men-of-war; the *Kent* and *Princess Anne*, and three Dutch, with two fireships, commanded by Sir Cloudesley Shovel, and that his Majesty, with the said yachts and men-of-war, was yesterday morning about 2 o'clock at anchor about 12 leagues E.S.E. off Yarmouth, and the wind being south it was believed his Majesty would land yesterday at Yarmouth, Southwold, or Orford.[15]

Unusually the King landed at Yarmouth, but this time there was a real threat, according to Narcissus Luttrell. 'As his Majesty lay at anchor off Yarmouth six French men of war, of about 50 guns each, appeared in sight and came within twice cannon shot, but the [sea] running high, and his Majesty convoyed by five great men of war, they sheered off.'[16]

On 10 October 1695 the Lords Justices had still not heard of the King's arrival and were in the usual suspense, until

King William's landing at Carrickfergus in June 1690, to lead his army in Ireland. He had travelled in the *Mary*, shown in the centre with the Royal Standard still flying. The *Henrietta* is to the left, carrying his brother-in-law Prince George of Denmark. (© National Maritime Museum, Greenwich, London, BHC0330)

'at 8 in the evening, Mr. Blathwayt's letter arrived giving an account of the King's landing at Margate that morning.' There were increasingly large celebrations on the King's return, with Engineer Richards being ordered to provide fireworks in 1696.

The peace

By the autumn of 1697 it was clear that the war had reached stalemate and peace negotiations were started. William was in Holland, as usual at that time of year, and on 2 September some of the yachts were sent there, though it was not expected that he would come back soon. The Treaty of Ryswick was signed on 10 September, effectively recognising William as legitimate sovereign, but still he did not leave and on 7 October it was reported that he would stay on, being 'resolved to see the articles of peace in some measure performed before he leaves Holland.'[17] On the 19th he sent for his yachts, while his government was ordering 'extremely fine' fireworks to celebrate his supposed victory. After the usual period of frustrated expectation in contrary winds, he was at Greenwich by 15 November to take part in a huge procession to London

The yachts returned to their usual round of carrying officials, aristocrats and diplomats. Apart from royal duties and long periods at anchor off Greenwich, the *Mary* took 'Lord Sandwich with his lady and retinue' on board in April bound for Calais. On 1 June they picked up the Dutch ambassador at Greenwich to take him home. In August they carried the Earl of Jersey from Greenwich to Calais. In September they sailed from Greenwich again to attend Sir Robert Rich, a Lord of the Admiralty, and bring him south from Lowestoft. In October they took the Duke of Grafton to Holland. In April 1699 they carried Sir Cloudesley Shovell to the *Swiftsure* at the Downs, and in July they brought Sir Joseph Williamson from Holland. In August the Earl of Sandwich was fetched back from Calais, and in November it was the turn of the Marquess of Tavistock. Diplomatic missions included bringing the King of Prussia's envoy over from Holland in May 1701, and a similar trip with the (Holy Roman) Emperor's envoy in November.[18]

Noblemen and courtiers still expected high-class transport. The confidence trickster, William Fuller, '[o]riginally a butcher's son, by education a coney-wool-cutter' tried to impress some English merchants in Holland in 1701, telling them that he was, 'a gentleman of the bed-chamber; that he was returning for England, and expected a King's yacht for his passage for which these gentlemen should be welcome.' No yacht appeared and 'instead of a King's cabin, our grandee was forced to make one with the merchants in the homely packet-boat.'[19]

Peter the Great

The most famous diplomatic mission of the age was the visit of Tsar Peter the Great of Russia to Holland and England to study shipbuilding. Plans had already begun when the King was negotiating the treaty, and the Tsar was 'still at Amsterdam, busying himself among the ship carpenters and blacksmiths working in the docks and is very inquisitive about navigation.'[20] The two monarchs had several cordial meetings, for William was keen to establish trade links with Russia, and on 12 October he offered him the gift of the yacht *Royal Transport* designed by the Marquess of Carmarthen, a well-connected amateur yacht designer and naval officer.

Early in 1698 Peter came to Deptford to study shipbuilding and navigation. He was put up in John Evelyn's house, Sayes Court, though he proved to be a very undesirable tenant and Evelyn's steward reported, 'There is a house full of people and right nasty.' The Tsar borrowed a yacht from a William Charlton of Greenwich, but his nautical skills were far from perfect. 'His Majesty, steering himself, run on board one of the bomb vessels and broke away figured rails and all that belonged to the head.' He also collided with the yacht *Henrietta*, causing her to be dry-docked with 'her joints shaken and her caulking loosed.'[21].

On 21 February the *Mary* yacht picked up the 'Czar of Muscovy' at Deptford, with orders to take him to Holland. There was a change of plan and she took him to Chatham instead 'to view the dock and the great ships'. They sailed again on the 3rd, but encountered gales and had to anchor off Sheerness and then Margate Roads. They finally left on the 25th, with the *York* and *Greenwich* men-of-war, and the *Henrietta* and *Expedition* yachts. Next day they anchored off Goree and the Tsar departed.[22]

Peacetime voyages

King William continued to visit the Netherlands during the peace, which caused some resentment among his English subjects, who felt they were being neglected. With no need to take part in the military campaigns, he started off slightly later, in June or July each year, and came back from October to December. On 17 July 1698 the *William and Mary* sailed from Greenwich with the King on board, in company with the *Mary*. They anchored at the Nore, then in Margate Roads, where they met Sir Cloudesley Shovell with nine men-of-war, to be joined by five more fifth and sixth rates – even in peacetime a large escort was required. They were unable to sail in the calm weather and the King went ashore several times before embarking at ten in the evening of the 29th in a west-northwest wind. They sighted the land early next morning, but it was calm and they used the flood tide for as long as possible before anchoring, and getting into Rotterdam early next morning.[23]

On 25 October Sir Cloudesley Shovell was again 'appointed to command the men-of-war that are to attend his Majesty in his return from Holland.' The government was keen to expedite the matter and on 1 November they expected 'every hour to hear that Sir Cloudesley Shovell is come into the Downs from Spithead, in order to his repairing to the coast of Holland with the men-of-war and yachts that are to attend his Majesty in his return.'

Shovell did indeed sail from the Downs that day in the *Swiftsure*, with the yachts *Mary*, *Fubbs*, *Katherine* and *Henrietta*. They arrived off Rotterdam three days later, but had to wait until the 19th when the *Mary* sent boats ashore to collect the King, but they came back without him – the wind had turned south-southwest and it was foggy, so he had presumably decided to wait for better weather. The wind oscillated between gales and calm, and it was 1 December before the King embarked, presumably on the *Fubbs* or *Katherine*, while the *Mary* took on the Lords of the Bedchamber. They set off, though there was still fog and little wind, and reached Margate Roads where the King and most of the passengers disembarked.

On 2 June 1699 the King went on board the *William and Mary* at Margate Roads, and they set sail accompanied by Shovell on the *Swiftsure* and the *Mary* yacht, among others. They had to anchor off the North Foreland to await a wind, but were off Rotterdam by the 5th. The process was reversed on 26 October, when the King boarded the same yacht at Brielle and went ashore at Margate the following day, with the yachts returning upriver to Greenwich as usual.[24] There was another voyage in 1700: it was reported, 'The king got well to Holland in 26 hours, and has lost the pain he complained of at his stomach.'[25]

Rooke at Copenhagen

In 1700 the *Mary* made a rare trip outside British, Irish and near-continental waters, still under the command of Sir

William of Orange rides off to the left after landing at Orange Polder in 1691, on the way to the war in Flanders. The Dutch fleet is in the background. (© National Maritime Museum, Greenwich, London, PAH5237)

George Rooke. He was to lead a squadron sent to Danish waters to co-operate with the powerful Swedish navy and prevent a general northern war, which might interfere with William's plans for the oncoming war with the French. Rooke joined the *Mary* at Greenwich on 30 April, but he was unable to sail downriver in her due to the easterly winds, so he was rowed by barge instead; the *Mary* joined him at the Downs two days later and was sent on missions to Calais, Dunkirk and back to Greenwich before rejoining on 12 May. He drew plans for his line of battle with his ten major warships, with the *Mary* as part of the auxiliary of the admiral's division, serving as tender to the flagship. The squadron arrived at Brielle on 15 May, meeting the *Katherine*, which carried messages back home. Rooke sailed in the 72-gun *Shrewsbury* with the *Mary* in attendance. By 17 June they were off Elsinore at the entrance to the Sound between Sweden and Denmark. On 4 July she was sent to carry Hugh Gregg, the English minister in Copenhagen, to warn the Swedish fort at Landskrona that the fleet was only going there for provisions and water, and not to be surprised by them. Rooke was on board the yacht on the 9th when he drafted an ultimatum to the Danes. That produced no result, so the admiral was on board again on the 16th to go close inshore as an English, a Swedish and two Dutch bomb ketches opened fire on Copenhagen. The *Mary*'s log recorded, 'The bombs played the last night till ten this morn.' But Rooke concluded, 'they did not do the execution that was intended', and he ordered a withdrawal. Later the *Mary* ferried troops around in support of a Swedish invasion. There was a compromise peace and eventually the squadron sailed.

They were in Dutch waters on 17 October when they encountered other yachts, including the *William and Mary*. The King boarded the latter and they set sail for England in 'a gentle gale and fair weather easterly.' But at eleven that night the wind veered to the southwest and the yachts had 'much ado' to reach Harwich next evening. They anchored in the harbour and according to the log of the *Mary*: 'The King went ashore with the lords of the bedchamber.'[26]

THE CHANGING ROLE

~2~
New Yachts

~ *William* and *Mary* ~

At the end of 1693 the *Fubbs* was fitted out to take the King to Holland but His Majesty, though satisfied, asked for a new yacht to be built in time for his return. Next month the Navy Board was ordered to prepare a draft after consulting with Greenville Collins. It was to be drawn by the Surveyor of the Navy, Edward Dummer, said by Pepys to be 'an ingenious young man, but said rarely to have handled a tool in his life'.[1] Normally the draft was the prerogative of the master shipwright of the yard concerned rather than the head of naval shipbuilding, but Dummer was determined to enforce higher standards of accuracy on the builders and often took on such work. The design was accepted by 13 February, which perhaps suggests he had been able to prepare much of it earlier, and it was estimated to cost £2060.[2] The new yacht was laid down at Chatham and was apparently nearing completion by 18 August, when the Admiralty accepted the opinion of the Navy Board that 'the new yacht building in the Chatham yard should be allowed the same complement of men as is established for the *Fubbs* yacht, namely 40 men in war at home and 30 in war abroad, and peace at home and abroad.'[3] But by the end of the month the Commissioner at Chatham had to report that, despite courtiers urging haste, 'The new yacht being built for the King will not be completed in time to bring him home.' It was to be launched on the next spring tide and could be gunned and provisioned, but it was not possible to complete two of the cabins and sails still had to be sent. It was to be launched around the 12th and Captain Sanderson was advised by the master shipwright to apply to 'the Wardrobe' for the furnishings of the cabins.[4] On the 7th the Admiralty wrote to the Navy Board.

> You have informed us by your letter of August 31st, that the new yacht building at Chatham for their Majesties' particular use, which we have named *William and Mary*, will be ready for launching about the 10th of this month.
>
> We direct you to cause her to be launched accordingly, fitted for Channel service, manned with her highest complement of men, and victualled now, and from time to time as needed, in the same manner as the other royal yachts are victualled, and entered and registered on the list of the Royal Navy by the same name.[5]

But it was reported that the 'figuring', executed as usual by the Fletcher family, was 'very badly done as it abused the King and deceived us all.' Perhaps the death of one of the brothers had affected the quality, or possibly a secret Jacobite took the

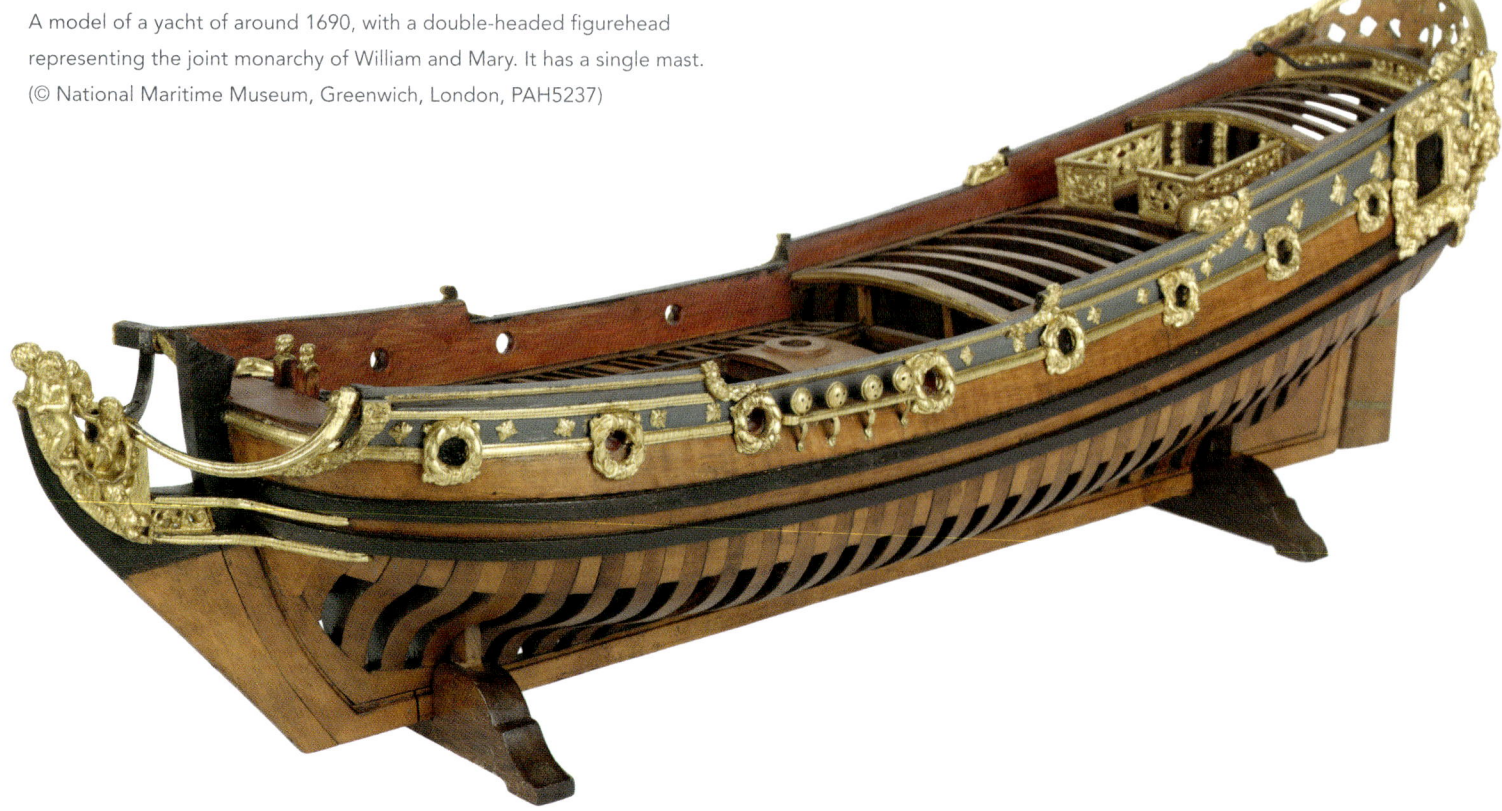

A model of a yacht of around 1690, with a double-headed figurehead representing the joint monarchy of William and Mary. It has a single mast. (© National Maritime Museum, Greenwich, London, PAH5237)

The *Carolina*, formerly the *Peregrine Galley*, by L de Man. The vessel is depicted in great detail, with decorations and men on decks and aloft, but no distinguished passenger who would be entitled to a flag. (© National Maritime Museum, Greenwich, London, BHC0979)

chance to make the King look less than attractive.[6] There were the usual problems with ballast: Robert Lee the master shipwright reported that there was no lead in store, but the Commissioner had hopes of contracting with Mr Slater 'at a lower rate than allowed the widow [Ann] Parsons.'[7] That was not the end of the problems. In July 1695 Sanderson reported that 'its mast is not taut enough to hoist the sail, and otherwise inconvenient.' The Navy Board ordered this to be remedied, but in 1702 it was sprung during a passage to Holland.[8] A more fundamental problem was with the ballast under the staterooms, for it had to be placed so tight that Captain Sanderson reported in February 1696, 'the lead ballast cannot be lifted without handscrews to clear the limbers so that the bilge water can be pumped out, this water being very offensive to all and causes tarnishing.' He asked for hand screws to solve the problem and a month later he asked for the floor to be veneered and oar ports cut for rowing. In May the King complained that the ship's boat was too small and asked for one 28ft long – that was easier to resolve.[9]

The *William and Mary* was the first true royal yacht after the revolution of 1688 and carried the King several times, as she had been designed to do. Though she was not large at 152 tons, she was used as a warship in the war of 1702–14 and repaired or rebuilt at Deptford in 1713/14, 1736/7 and 1764/5.

The *Royal Transport*

Peregrine Osborne was a naval officer and talented amateur ship designer, who confusingly took on different titles as his father, leading government minister under Charles and William, was promoted in the peerage. Peregrine was the Earl of Danby when he drew the plans for a very strange vessel with a 'wasp-waisted' midship section. It would have been very difficult to construct and there is no evidence that anyone ever tried. In 1694 he built the *Bridget Galley*, which was regarded as a good sailer but too expensive for naval use, so it became a privateer. His next effort was a less radical but still experimental design, which was to be built by Lee at Chatham. The most innovative feature of the hull was that the 'floors' of the lowest part drooped down so that they were level with the keel – perhaps, like the wasp-waist of the 'maggot', intended to increase draught as the ship heeled. It used the 'schooner' rig, fore-and-aft but distinguished from the much more common ketch in that the mainmast was taller than the foremast. Indeed, the term 'schooner' was not yet coined until it originated in New England a few years later, and the type did not appear on the Navy Lists until 1764.[10]

The ship was launched on 11 December 1695 and began wages and victuals on 20 January 1696. Her crew included 'three trumpeters belonging to the Marquis of Carmarthen'. On 11 March the Duke of Shrewsbury, one of the secretaries of state, asked that 'the Sixth Rate Frigate built by the directions of the Marquis of Carmarthen at Chatham, be called the *Royal Transport*.'[11] Despite her name, which accurately signified her function, she was registered on the Navy List as a sixth rate – at 90ft long in the main deck, 23ft 6in broad and measured at 220 tons, she was presumably considered too big to pass for a yacht and she carried eighteen guns to emphasise the point. Nevertheless, on 1 April 1696 the Admiralty ordered 'the *Royal Transport*, now at Deptford, to be victualled from time to time as needed, in the same manner as the Yachts are victualled.'[12] Carmarthen was still in charge of the project, with the support of the King. On 6 June the Admiralty ordered, 'In pursuance of his Majesty's pleasure signified to this board we direct you to cause such alterations to be made to the *Royal Transport* as are desired by the Marquis of Carmarthen.' This annoyed the Lords of the Admiralty, who complained: 'building his Majesty's ships according to the fancy and humour or particular persons will break through those wholesome rules and methods which have been found absolutely necessary for the well performance of his Majesty's service.'[13] Her interior was well fitted, as indicated in a letter to the Navy Board in 1696:

You advised by your letter of 1 October that the officers at the Deptford yard have been ordered to survey the great cabin of the Royal Transport. They find that the Japanning is very well done and agree with what the Marquis of at Carmarthen agreed with Mr Gumley for a cost of £250, with which you have no objection.[14]

Her speed was universally praised, she was described as the 'best sailer' and the 'fastest and best yacht [*sic*] then in England'. Carmarthen was not noted for his modesty and wrote that a 'small craft of my invention was designed not only for comfort and utility but also that it should be faster and more powerful than other ships exceeding it in size.'[15] She made several one-day passages across the North Sea under Carmarthen himself and Captain William Ripley, who was even more erratic than the marquess, and it was suggested that he cut away some of her timbers to improve the accommodation.[16] She carried distinguished passengers such as the Earl of Portland.

By November 1696 it had been decided to turn her into a fighting ship as her rate, if not her name, implied. The Admiralty ordered, 'As soon as the *Royal Transport* arrives at the Nore, you are to have her surveyed and let us know what alteration must be made to put her in condition as a frigate.' But going on board, the surveyors found that she was 'so full of noblemen's goods we could not go forward and aft in her.'[17] On 16 April 1697 there was an order 'urgently requiring the *Royal Transport* at Deptford to be put in a condition for sea in order to accompany the King to Holland.'[18] That does not necessarily contradict the previous order – she was presumably intended as an escort for the yachts.

In any case, a very different fate was in store for the ship. In January 1698 Peter the Great, Tsar of Russia, arrived in Deptford to study shipbuilding. The fame of the *Royal Transport* might have been part of the attraction, and in the previous October a letter indicated, 'His Majesty being desirous to gratify the Czar of Muscovy in what may be most acceptable to him, has thought fit to present him with the *Royal Transport*'. Carmarthen agreed, but asked, 'having been the contriver of that vessel, he may be permitted to put her in a better sailing condition that he supposes her to be at present, by the alteration of her foremast and otherwise'.[19] The handover of the ship was not without embarrassment. Captain Salmon Morrice had replaced Ripley after a court martial, but was surprised on 11 March when his predecessor appeared with instructions to take over the ship. He rode to London to have this confirmed and was given command of another ship. She sailed a month later with Ripley in command and Peter on board, but he did not go on to Archangel as originally planned, disembarking at Hellevoetsluis.[20] The *Royal Transport* continued the journey and was a feature of the Russian Navy until 1715.

The *Peregrine Galley*

A new type of fast warship was developed in the 1670s. The galley-frigate was intended to counter the Barbary corsairs for, as Pepys observed, 'Our seas require stronger and therefore heavier ships, which spoils their sailing, and therefore not the best in calm seas, and therefore we should have ships built to be employed only in the straits' – ie the Mediterranean. As its name implies, the new type was intended to operate under both sail and oars to match Barbary vessels such as the xebec, but it performed better under sail. As captain of the *Charles Galley* in 1688, Jeremy Roch regarded her as 'the best sailor in the fleet'. 'I had a mind to try how the galley would row, so I ordered the sails to be hauled up and ran out all our oars being 42 and put three men on each oar and found we could row three miles an hour'.[21] There is no evidence that they were rowed much after that and indeed Roch did not try it on occasions when his ship was becalmed. Charles abandoned his possession of Tangier in 1684, so there was no base for the galley frigates and they tended to serve more in northern waters.

By the end of the century the Marquess of Carmarthen reckoned he was owed £9000 in pensions granted to him by Charles II. William's cash-strapped government found an extraordinary way to compensate him – he was allowed to build his own yacht at government expense. She was to be 86ft 6in long and 22ft 10in broad, measuring at 197 tons, nearly a third bigger than the *William and Mary* at 153 tons. She was to be constructed at Chatham by William Lee who had already built the *Royal Transport*, but early in September 1699 Carmarthen visited the yard and found that Lee had been transferred to Sheerness. The frames for the ship had already been cut out and Carmarthen asked that they be sent down the Medway to join their builder.[22] This was agreed and less than a month later, on 6 October, Lee was asking for 'orders to launch the vessel built for the Marquis of Carmarthen'. But if the assembly of the hull was particularly rapid, the fitting out was unduly slow. In August 1700 Carmarthen, no more economical than he was modest, asked for walnut veneers in his cabin and in September directions for carving had not yet been issued. Perhaps he intended her as a yacht, but with potential for twenty guns and high speed she could also have served as a privateer.

She was still known as a 'galley' because there was some provision for rowing with sweeps put through ports in the lower hull, though she was mainly a sailing vessel, like the others of similar denomination. As a result, she was given the name *Peregrine Galley* after her designer and prospective owner. She was slightly narrower than other ships of the times, and had much less draught and sharp ends, presumably to make her easier to row. She had a bell-shaped keel which spread outwards in cross section, intended to increase the draught of the ship when heeling on the same principle of the winged keel of the 1980s.[23] She apparently attracted naval attention while she was still on the stocks, was taken over and was in naval pay from 20 September, though she was still not ready to attend the King in Holland; Carmarthen disobeyed orders to remove her to Deptford for final fitting.

William Sanderson was not entirely happy with the fitting of her waist when he took over her command from Carmarthen in 1702 and he complained in April:

> … her gunwale under the planksheer not being filled, the water comes down in to the boades [?]; the fife rail likewise is supported by a great number of iron stanchions, all which I desire may be taken away; and pieces of iron bores let in to support the same; and that I may have a convenient gallows made to set my boats on; for when the boat is upon the decks, if it should blow hard so that we cannot get her out, we cannot short in upon our cable through the ship, and our lives are at stake. … and that several other small things may be done, that may make the ship safe in the sea.[24]

Her early rigging is something of an enigma. There is no picture of her in those days and a contemporary model lacks rigging, but shows channels for shrouds for two masts, a large one amidships with six deadeyes, and another well forward with five. This is different from the now traditional ketch rig of the yachts in that it has a foremast rather than a mizzen. This suggests some form of fore-and-aft rig, which was unusual for a vessel of that size at that time.[25] She was apparently re-rigged with square main and topsail yards on her fore and main, and with a square sail hung from the bowsprit. That did not prove satisfactory and Captain Sanderson pointed out in April 'that my topmasts may be made to keep that when it blows hard, I may have no more mast than sail by shortening the same'. By December he knew more of her sailing qualities: 'relating to altering the *Peregrine Galley*'s masts and sails, I am of opinion that if she was masted and sailed in proportion to her body and the accustomed rules of the navy, she would not only sail better, but be safer in the sea.'[26] He complained of 'the said galley's want of after sails, which could not be obtained in her, and without placing a mizzen mast'. Though she did not fit with the regular 'establishment of the navy', the officials of the Navy Board and dockyards agreed that she should 'entirely be masted after the frigate fashion.' They proposed a new rig with slightly shorter lower main and foremasts but with short topgallants added above to increase the overall height, and with the addition of topgallant yards and sails. She was indeed to have a mizzen with a topmast and yard, and a crossjack to spread the foot of the mizzen topsail. Her bowsprit was also to have a topmast added, with its own yard. As a result she would

to be able to carry eleven sails rather than five, in a more or less conventional ship rig.²⁷

Like the *Royal Transport*, the *Peregrine Galley* also proved a good sailer with her modified rigging, and Carmarthen was a victim of his own success. In 1703, after the death of William, he applied again to have the galley restored to him, but was told that the Admiralty Council 'considered that so fine a vessel should not be separated from the fleet; and the Queen kept the vessel for her own service.'²⁸ Classed as a sixth rate warship of twenty guns, she actually served as a VIP transport during the early 1700s. In 1714 she was chosen to bring King George I to England for the first time and was converted to a fully-fledged royal yacht with the name of *Carolina* at Deptford in 1716. She had not been designed for square rig in the first place, nor as a

Princess Charlotte

The next ship to be officially registered as a royal yacht was the *Princess Charlotte*, built by Joseph Allin at Deptford in 1711 and measured at 155 tons. On 5 March 1711, one William Boswell bought a firkin of soap to grease the ways for her launch, which took place five days later. By 25 April she had been hauled out of the dock, presumably having been graved, and orders were requested for her furniture and stores. These were quite elaborate and the King's bedchamber was furnished with a 'green camblet stand bed with curtain and valance' which was 'worn and much soiled by 1727', while the stateroom had a feather bed and the other cabins had to make do with flock beds. But in the first instance she was used mainly to pay ships at anchor in the Medway and Nore. She was re-rigged as a ketch in 1737, renamed *Augusta* in 1761 and rebuilt, essentially as a new ship, at Deptford in 1771.

The older yachts

The transfer of most of the activity of the yachts from the Thames Estuary and the English Channel to the stormy North Sea had perhaps exposed some of the weakness of the Stuart vessels. A low hull caused the ship to make less leeway, but in rough weather it could make the deck very wet. In 1713 Captain Desborough complained that the *Fubbs* 'carries much water, thereon being very inconvenient', and the Navy Board ordered her deck to be raised by four inches during a repair, 'which may now very well be done, the knees being taken off in order to shift the clamps, and that the breadth will also bear it.'[29] The *Cleveland* on the other hand was found to be crank, that is, 'incapable of carrying sail without being exposed to the danger of oversetting.' She had been repaired at Deptford in 1712 and was filled with 4 tons of iron shot after new wooden riders had been fitted in the hold, but when the provisions on top of that were used up she proved 'not so stiff as before'. The yard officers suggested using 6 tons of lead ballast, which would remedy the fault and also allow more room in the hold. The *Henrietta* was also reported to be 'very tender sided – her iron ballast lying high helps considerable to that misfortune'. Deptford asked for 4 tons of old lead in replacement. And of the *Mary* it was complained that, 'We find that her work between wind and water [ie close to the waterline] to be defective and cannot be made good bringing her into the dry dock', though that might have been the result of neglect or a previous repair, rather than her original building.[30] Nevertheless, these were essentially the yachts which would carry kings and nobles well into the century, after a new regime had taken control.

The Peregrine Galley is in the centre of this picture by Jan Griffier the Elder, flying the Royal Standard. Greenwich can be seen in the background. (© National Maritime Museum, Greenwich, London, BHC1821)

royal yacht. Thus, by accident, she moved the royal yachts away from the small fore-and-aft rigged vessels of the past, and introduced them to ship rig, with three masts, all square rigged.

3
Marlborough at Sea

William of Orange died on 8 March 1702 and was succeeded by his sister-in-law. Queen Anne did not travel much, and unlike her predecessor and successors she had no territory in Europe to rule, but again the yachts were used in supporting a war in Flanders, which began on 23 April. Her great general, the Duke of Marlborough, took on the leadership of the allied forces. Since campaigning was seasonal, Marlborough returned to London most winters to keep his finger on the pulse of politics and maintain his relations with the Queen. He seems to have favoured the *Peregrine Galley*, though this is difficult to substantiate as his letters are rarely specific about vessels, and logs from the period are few and far between. He certainly used her in April 1706 and in April and June 1709.[1]

In the meantime, Sir George Rooke used yachts again with his fleet in a maritime operation. On 20 May he asked for the *Fubbs* to be cleaned and tallowed at Deptford. He sailed in her on the 26th, taking £8000 contingent money to the fleet at the Nore, where he boarded his flagship, the *Royal Sovereign*. The *Isabella* was ordered 'to attend his Grace the Duke of Ormonde for the present expedition' – he was commander of the land forces in a project to capture Cadiz. They were off the port on 11 August when the *Fubbs* and two other ships were ordered to land 120 grenadiers 'for intelligence'. The expedition failed and they withdrew, but they sailed to Vigo in the north of the country where they entered the harbour and captured six fine French warships and a quantity of treasure. The *Fubbs* was back in home waters by 12 August, with five 'great ships'. She carried Rooke up to Greenwich on 16 and 17 November, and four days later it was hailed in the House of Commons as 'a most glorious expedition'.[2]

Marlborough in Flanders

At the end of his first campaign in Flanders, Marlborough had a dangerous moment in a yacht, albeit well inland. In November 1702 he was in a Dutch yacht being towed along the Maas between Maastricht and The Hague, visiting several towns along the way. They became separated from another boat and their cavalry escort and fell foul of a group of French raiders, led by an Irish lieutenant who had deserted the Dutch service.

> Between eleven and twelve at night a party of thirty-five men of the garrison of Guelder, who lay skulking on the banks near three leagues on the side Venloo, having by surprise seized the rope with which the boat was drawn, and hauled it on shore, they immediately made a discharge of their small arms upon us, and then threw several grenades, with which some of our people being wounded, the party entered and seized the boat, and having examined the several passports without known my Lord Marlborough, they afterwards searched the trunks and baggage, from whence they took what plate they could find, and likewise made our part of foot prisoners, and about five in the morning retired with their booty, leaving their Excellencies to continue their voyage.[3]

Marlborough only escaped because one of his attendants handed him a spare pass. After that he usually preferred to cross the seas with a naval escort. As late as November 1711 he was hoping for 'the yachts and convoy to come over for me'.[4]

Marlborough's comments on the voyages across the North Sea were mostly expressing frustration about delays due to wind and weather. He returned late during the winter of 1703/4 and in February he complained, 'The wind being about on Friday last to the NE, I was in hopes to have made my passage to England before the packet'. He left The Hague by torchlight and planned to embark at Oranje Polder, 'but before I reached the waterside the wind was come about again to the NW, so the yacht could not get down the river'. He crossed over to Brielle on the south bank of the Maas, only to find that the transports that were supposed to accompany him had already sailed. The *Mary* yacht picked him up at seven in the morning of 12 February, but promptly ran aground due to the negligence of the pilot. Marlborough, impatient as ever, disembarked and took the regular packet boat – but the *Mary* got off soon afterwards,[5] and that April he had 'a very easy passage, notwithstanding the wind was for the most part easterly.' They were becalmed off the Maas and the tide was unfavourable but Marlborough was rowed ashore.[6]

At the end of March 1705 he was delayed at Harwich:

> I went on board yesterday noon, with the wind at SW and a fair prospect of a good voyage; but when we came to join the convoy in Ousley Bay, the weather then being hazy, the pilots would not venture out with the men-of-war, so that we were obliged to come to an anchor with all of the transports, expecting to pursue our voyage this morning; but at five o'clock, the wind coming about to the NE, and a strong gale with very foul weather, forced us back.[7]

His most frustrating time was in the spring of 1707. He arrived at Margate on 22 March to find the wind contrary, though the sailors offered hope of 'a favourable change tomorrow'. But he was there for more than a week, until the squadron set sail on the 30th, but after about twenty miles it was beaten back by 'a north-east wind, which still continues'. He boarded the *Cleveland* on Wednesday, 2 April. The wind was still unfavourable at northeast and east, but they progressed across out of the estuary and across the North Sea by 'tiding', or anchoring every time the tide turned against them. On Saturday 5th the wind changed to southwest and they arrived off Goree and then went into Hellevoetsluis. Marlborough was worried that the delay had allowed the enemy time to organise. His duties were as much diplomatic as military in holding a diverse coalition together, and on his arrival he was obliged to write to numerous allies in the language of diplomacy, apologising for delays due to 'les vents contraire'.[8] He was much happier a year later when he wrote 'how lucky I was in making my passage hither [The Hague] the next night after I left London.'[9]

The end of the war

Marlborough won great victories at Blenheim in 1704, Ramillies in 1706, Oudenarde in 1708 and Malplaquet in 1709. The war began to wind down after 1710 and Marlborough's influence at court declined drastically, but he continued to voyage across the North Sea. On 18 February 1710 he boarded the *Peregrine Galley* at Greenwich and disembarked at Goree on the 22nd. He boarded the *Peregrine* again at Hellevoetsluis on 22 December. The ship and her escorts dropped down to Goree and met a thick fog. It was calm the next morning, but a fresh gale sprang up in the afternoon and allowed them to put to sea for a few miles. Christmas Day saw them at anchor again, but they sailed all night with an 'easy sail' and they arrived in Southwold Bay at 3pm to land the duke. Between these voyages, the *Peregrine Galley* and the other yachts carried ambassadors, and Marlborough's very effective ally, Prince Eugene of Savoy.

The war was becoming increasingly unpopular and there were riots in the country. In October 1710 a general election saw the defeat of Marlborough's Whigs and put the Tory party in power. Sarah, Duchess of Marlborough, had lost her influence with Queen Anne, now a very sad figure after the loss of all her children. Marlborough himself would continue to campaign, but with much reduced power. He boarded again at Harwich on 19 February 1711 and was put ashore at Rotterdam three days later. He had some success in penetrating the *ne plus ultra* lines which would have allowed the allies to advance into France, but the politicians and public were not interested. He had no role in the peace negotiations which were taking place though he looked forward to 'putting an end to this ruinous war'.

On 7 November (English style) he wrote from The Hague:

The yachts and convoy came very luckily on Friday last into the Meuse and Goree, for the westerly winds have blown so very violently ever since that they might have been in some danger on the coast. This stormy weather has hindered my leaving this place yesterday as I intended, having ordered the yachts round to Goree, but I resolved to go to the Brill tomorrow to be nearer at hand to make use of the very first opportunity of wind and weather to put to sea.

He boarded on 15 November and they sailed in convoy with the *Mary*, *William and Mary*, *Fubbs* and *Cleveland*. With a favourable south-by-east wind they sailed up the Thames to arrive at Greenwich on the 17th. It was his last voyage and service to the Crown. The War of Spanish Succession ended with the Treaty of Utrecht, which was signed on 26 June 1714. Queen Anne died a few weeks later, as did Louis XIV of France in 1715. The role of the yachts would change again as they helped to establish a new regime, in a generally more peaceful world.

John Churchill, 1st Duke of Marlborough, by or after Sir Godfrey Kneller. (Royal Collection Trust / © Her Majesty Queen Elizabeth II 2022, 404014)

Part III

The Hanover Connection

1
The Coming of King George

The *Peregrine Galley* was undergoing routine maintenance at Deptford at the end of July 1714 – her bowsprit was rotten and she was hauled out to the sheer hulk offshore to have it replaced. Momentous but not unexpected news was received on 1 August. Queen Anne, the last of the Stuart sovereigns, had died. She had had fourteen pregnancies, but none of her children had survived beyond the age of fourteen. There was a possibility that the political classes might turn to the Jacobite claimant who called himself James III, otherwise the Old Pretender. Some of his leading supporters considered proclaiming him at the Royal Exchange, until 'Lord Bolingbroke said that all our throats would be cut'.[1] The Jacobites would remain strong in places for another thirty years, but all of the Whig party and half of the Tories were determined to keep them out.

Instead, they turned to the next nearest heir to the throne, George, Elector of Hanover – so called because he had a vote in the college which elected the Holy Roman Emperor. After a century which saw civil war, revolution and bitter party strife, the country craved stability, even at the price of rule by an unattractive and alien king, and an aristocratic oligarchy. So in front of a great crowd, a herald proclaimed from St James's Palace, 'That the high and mighty Prince George Elector of Brunswick Luneburg is now by the death of our late sovereign of happy memory, become our only lawful and rightful liege lord, George by the Grace of God King of Great Britain, France and Ireland.' The crew of the *Peregrine Galley* played their part in recognising the transfer, firing a seven-gun salute in the afternoon 'for the proclamation of King George'.

The *Peregrine*, though not formally registered as a royal yacht, was by far the most suitable vessel to bring the King over. The other yachts were much smaller and the regular warships were too spartan to accommodate a king. George was in no hurry to get to his new realm. His agent and his supporters in Britain secured the regime with control of Parliament and the army, and George remained in Hanover for several weeks to put the Electorate in order. The *Peregrine Galley*'s routine was undisturbed for several days and the log only recorded the weather – 'some rain' on the 9th, 'fresh gales with squalls of wind and rain' on the 10th. Then on the evening of the 13th, amid 'fresh gales', 'the things belonging to the [Board of] Green Cloth came aboard and the King's Household family goods.' Next morning they slipped their moorings, sailed downriver and anchored overnight in the King's Channel in the Thames Estuary. They had a fast passage across the North Sea and sighted Goree just after noon on the 15th. They tacked to await a pilot, who finally appeared at ten in the evening. They got into Hellevoetsluis at slack water next morning and began to tallow and rosin the ship's sides, when Admiral Lord Berkeley appeared in the *Mary* yacht with five men-of-war. He was a staunch Whig and supporter of the House of Hanover, though later described as:

> a man of great family and great quality, rough, proud, hard and obstinate, with excellent good natural parts, but so uncultivated that he was totally ignorant of every branch of knowledge but his own profession. He was haughty and tyrannical, but honourable and gallant, observant of his word, but equally incapable of flattering a prince, bending to a minister or lying to anyone he had to deal with.[2]

He flew his flag from the *Monck*, with the new Prince of Wales allocated the *Mary*, and Lord Clarendon and the Prince of Hainault the *Henrietta*. Three more ships, the *Centurion*, *Windsor* and *Weymouth*

King George I as drawn by Kneller. (Royal Collection Trust / © Her Majesty Queen Elizabeth II 2022, 913253)

George I arriving in the Thames, presumably for the first time. The *Peregrine Galley* is on the centre right, and to the right Tilbury Fort is firing a salute. Other yachts present include the *Henrietta, Katherine, Cleveland, Mary and Charlotte*. By Peter Monamy. (Richard Green Gallery)

arrived a few days later, for twelve warships and a sloop were sent to Holland 'under the command of the Earl of Berkeley to attend His Majesty from thence to Great Britain.'[3] They were to escort the six royal yachts.

The ships remained at Hellevoetsluis for nearly a month before moving out on 14 September. King George I came on board the *Peregrine Galley* at Orange Polder on the 16th and the *Peregrine Galley*'s log records, 'We put out all our silk colours before he came on board, at his coming on board, hoisted the Royal Standard at main topmast, Lord High Admiral's flag at fore, [?] at the mizzen.' The last part is difficult to interpret but presumably means the Union flag. This is borne out by the log of the *Charlotte*, which records, 'the Standard and Admiralty flag and union flag flying on board the *Peregrine*.'[4] The ships sailed from Dutch waters on the 16th. The log mentions thirteen men-of-war and a sloop, which is one more than in the list book, as well as five Dutch warships, all of which saluted twice.

They sailed in a 'a fine gale' from the east-northeast on a course of west-northwest, and made a fast passage, sighting the English coast at just north of Southwold at six next morning. Then, 'we run up the back of the Shipwash with all sails we could make up the King's Channel', until they reached the Nore by six in the evening of the 17th and were saluted by the ships anchored there. The job of the major warships, to escort the yachts, was done now that they were out of the open seas and they fired a salute while the yachts proceeded up river. Admiral Berkeley transferred his flag from the *Monck* to the sloop *Jamaica* to follow them up the Thames: 'at 5 the admiral and all the fleet saluted His Majesty. Ditto the earl of Berkeley hoisted his flag on board of us … the admiral went in our boat, we saluted him with nine guns.' By midnight there was a thick fog and the ships anchored at the lower end of the Hope. It cleared by ten the next morning and the yachts 'turned up at Gravesend' opposite Tilbury, where there was a collision with a smack from Barking, which hit the yacht's quarter but did little harm. A salute was fired by Tilbury Fort, and they got far as Long Reach by 4pm, when the King went into a barge to take him to Greenwich, where he would be officially received.

The nobility were instructed to attend in Greenwich Park the next day and their coaches were marshalled to form a procession with the juniors at the head and the King behind. At Southwark they were met by the Lord Mayor of London and his aldermen and sheriffs, and proceeded towards St James's Palace accompanied by kettledrums and trumpets. It was probably the King's greatest moment of popularity and triumph, for apart from the Jacobite issue he was unpopular and was personally reserved and unattractive; events surrounding his coronation in October were disrupted by riots.

After the delivery of the King, the *Peregrine Galley* reverted to routine. Back at Deptford, she was hauled up on the ways and breamed on her sides. She was converted to a fully-fledged royal yacht with the name of *Carolina* at Deptford in 1716, with a layout that would become standard. She had two cabins aft with the royal bedroom above and a less luxurious one below. Forward of that, Charles's yachts had had a royal dining room with an open deck above, protected by a canvas awning in warm weather or rain above. This structure became permanent, but was still known as the awning deck. The crew's and officers' quarters were situated forward of the mainmast.

Neither George I nor George II ever became fluent in English. In true Hanoverian fashion they disliked each other intensely, but they had their love of their birthplace in common – they were true rulers there, not dominated by the fractious British politicians. They made sure that, war permitting, they had a sojourn of several months in the Electorate every two or three years. Both kings, however, had to observe the constitutional niceties, and were aware that they had to be close to London while Parliament was sitting, usually in the winter and spring. They could leave as soon as the session was over, sometimes very soon after, three days later in 1729 and two days in 1732 and 1735. The yachts had to be ready as the parliamentary session drew to a close. George I's intentions to visit Hanover were often frustrated by events: in 1715, in 1717 and 1718 because of political uncertainty. In 1721/2 there was the South Sea Bubble crisis and the threat of another Jacobite invasion. Visits in 1724 and 1726 were prevented by fears for the King's ill health.

Therefore George visited Hanover in 1716, 1719, 1720, 1723, 1725 and 1727. He stayed for five or six months, but often remained late enough to encounter bad weather, especially in 1725/6. But in other ways, as in many naval matters, the reign of William III set precedents which were followed throughout the eighteenth century. In the case of the King's voyages, these included carrying him in the largest yacht available – the *Royal Caroline*. Apart from the royal suite, accommodation was very cramped and spartan, so other yachts were needed to carry the high officials in suitable style. Invariably there were five of these, including the *William and Mary*, *Fubbs* and *Mary*, and either the *Katherine* or *Henrietta*. The *Fubbs* carried back the Duke of Montrose, the King's secretary in 1719, Earl Stanhope in 1720 and Lord Townshend in 1723 and 1725/6.

They were provided with a strong naval escort under the command of a distinguished admiral. More than a quarter of a century after the overthrow of James, loyalty was clearly a factor in the choice, and the admirals were all staunch anti-Jacobites. Matthew Aylmer who led the group in 1716 was an Irish Protestant, who also played a part in William's favour in the revolution of 1688. To his supporters he had 'a very good head' and was very zealous of the liberties of the people'; to his enemies he was a 'swearing idle fellow'.[5] Sir John Jennings, who led in 1720, was a strong Whig who had been out of political favour in the latter years of Queen Anne's reign. Sir John Norris was from another Irish Protestant family and was Shovell's protégé and Aylmer's son-in-law. He led fleets into the Baltic during the summers of 1716–21 and commanded the royal voyages in 1723 and 1725. George I invariably left government in the hands of the Lords Justices – his wife Sofia was not available, being imprisoned by her husband in Hanover.

The strength of the escort varied. For the outward voyage in 1716 it consisted of two two-decker 50-gun ships, three 40s, a 36-gun ship and a single-decked 20. This was much reduced for the return voyage in 1719, with only two 40s and two 20s. It was much stronger for the outward voyage the following year, with the 60-gun *Windsor* – the largest ship ever used on this service – plus a 50, a 40 and a 20. A projected voyage of 1722 was to be escorted by five 50-gun ships, but that never happened. In 1723 there were only three ships, the *Colchester* and *Leopard* of 50 guns and the 20-gun *Port Mahon*, for both the outward and homeward voyages; but 50s were high-sided and ungainly ships and the two of them were wind-bound in Holland and did not return until the following year. Perhaps that is why the outward escort in 1725 consisted of a 40, three 20s, and three tiny sloops of ten and eight guns each. The 50-gun *Leopard* was back for the return voyage, along with two 20s including the *Port Mahon*, which was a regular on these voyages, and three sloops. A similar party brought the King back.[6]

The passage out

The King usually embarked at Greenwich. In 1723 the yachts departed at seven in the evening of Monday, 3 June, but only had the tide to take them just past Woolwich that night; next day they made Gravesend by the same method. Next morning they had a fair wind and joined the escorts at the Nore by noon. They hoped to be off Holland by Thursday, but the winds were either light or easterly and they were still off the English coast at Orfordness. After Aldborough, strong southerly and westerly winds allowed them to begin the crossing, but it did not last and it took yet another day to reach Hellevoetsluis.[7]

They had a slow passage again in 1725, weighing from Greenwich in the evening of Thursday, 3 June, and tiding to Long Reach, then to Gravesend by Friday afternoon, when a fresh wind from the west allowed them to sail against the tide and join Sir John Norris and his escort at the Nore, They weighed on Saturday morning with a fair wind but it soon turned easterly so they had to tack against it until Sunday noon, when they saw the land. The wind was now in the southeast with thunder, lightning and squalls, and they anchored until break of day. They sailed over the Hinder at six in the morning and Norris sent the sixth-rate *Rose* back to inform the Lords Justices that the King had arrived and they could open their commission.[8]

Hellevoetsluis was the chosen point on the Continent for the King's trips, rather than his own Hanoverian port of Stade. Among other things, it allowed dealings with allies in the Netherlands and Germany, a point which was not always appreciated by the British public, and particularly the opposition politicians. According to Thomas Nugent in 1756 Hellevoetsluis was:

The *Carolina* carrying the King in a violent storm which drove it into Rye in 1726. (© National Maritime Museum, Greenwich, London, PAG6967

… a small sea-port town … This place is but small, consisting only of a handsome quay and two or three diminutive streets; but it is very well fortified, and esteemed the best harbour in the country. Here are generally some of the states' largest men of war, which lie at the end of the town. The English packet boat sets off from hence for Harwich twice a week, that is Wednesdays and Saturdays. Notwithstanding it is one of the most convenient places to embark at, and has so fine a harbour, it has but very little trade, the merchant-men choosing to go higher up the river before they unload.[9]

The route via Hellevoetsluis involved a 300-mile land journey to Hanover, though the King might travel the first part in a Dutch yacht, as he did in 1725 – he had already taken one in the opposite direction in 1720. After disembarking the King, the yachts went back to Greenwich and the warships resumed their normal duties.

∼ The return voyage ∽

The yachts were sent back to Hellevoetsluis in good time for the King's return and they usually had to wait for him. In November 1723 Admiral Norris was there awaiting the sudden arrival of His Majesty and ordered his captains, 'to keep all your upper tier of guns in constant readiness, with their full loading of powder, to salute his Majesty upon embarking, and to cause them to be fired when I have hoisted an English ensign at my mizzen peak and fired three guns'.[10] He finally arrived, embarked on 27 December and was saluted by the usual seven guns from each yacht.

They also had to wait for suitable wind and weather, which was not easy to predict. In November 1720 the *Royal Caroline* hauled out from the pier at Hellevoetsluis, presumably to make room for a Dutch yacht which arrived carrying the King. His Majesty went ashore, then was rowed out to the yacht next day. The sailed, but were obliged to anchor off the Goree lighthouse and then turned back in 'fresh gales' from the southwest. The King went ashore and reboarded two days later. By that time the wind was in the southeast and they soon reached North Foreland and then Margate, where the King disembarked.[11]

A ship-rigged royal yacht, presumably the *Carolina*, is shown stern-on flying the Royal Standard, so presumably it represents the return of George I or II from Hanover. (Royal Collection Trust / © Her Majesty Queen Elizabeth II 2022, 402429)

One of the worst voyages was early in 1726, the King's departure having been delayed. An easterly wind was highly desirable, but not if it was too strong. The yachts set sail on 1 January, only to find a 'hard gale of wind and very thick weather, with snow and sleet', and the *Fubbs* was obliged to take two reefs in her main topsail. They got over the Hinder bank but were obliged to heave to. They sailed for a few hours after handing the main topsail but were obliged to lay to again in 'great seas'. Somehow the squadron remained substantially intact and they obeyed signals to wear to the north and then to the south, while they took in two reefs in the mainsail. There were still '[h]ard gales of wind with much snow and rain' by the afternoon of the 3rd and they laid up again under the mizzen alone, but at least the passage was mercifully fast and they sighted the South Foreland light near Dover at six in the morning. They sailed for Rye Bay where they were sheltered from the easterly winds by the promontory of Dungeness, and the King disembarked there on the 4th; the *Fubbs*'s passenger, Lord Townshend, soon followed with his wife.[12]

Apart from that, and in 1719 when he sailed all the way up to Greenwich, the King mostly landed at Margate, a town which was barely noticed by Defoe in the 1720s but was developing as a spa by 1751. It was 'noted for shipping vast quantities of corn … for London; and has a salt-water bath at the post-house, which has performed great cures in nervous and paralytic cases, and numbness of the limbs.'[13] It had a small harbour in which distinguished travellers could board or disembark from boats to take them to and from the yachts. These were anchored offshore in an area protected from the north by Margate Sand or 'Hook'.

George left St James's Palace for Hanover again on 3 June 1727 and boarded the *Royal Caroline* at Greenwich, where the courtiers who were not going on the trip dined on board. Due to contrary winds he was not able to sail until the 5th, but he was off the Dutch coast by eight the next morning. He transferred to a Dutch yacht and landed at Schoonhoven to begin his usual carriage ride. After stops at Varth, where he had a miserable dinner, and Apeldoorn, he spent the night at Deaden. They set off at seven the next morning, but the King had spent a bad night with stomach pains. He had the carriage stopped and he was given smelling salts. A surgeon was called, who diagnosed a stroke. He went on to Osnabruck on the borders of Hanover but he died on the night of 10/11 June 1727. There were proposals to bring his body back to London, which would have provided another role for the yachts, but instead he was buried in his birthplace, while his regime continued in Britain in the hands of his son.[14]

2

The Yachts at Deptford and Greenwich

The role of Deptford Dockyard

Deptford Dockyard was developed by Henry VIII, round a wet dock which was intended to keep his largest ships, including the *Mary Rose*, afloat in winter. It was close to London, and even closer to Henry's favourite jousting ground at Placentia, Greenwich. It was well up the River Thames, which protected it from enemy incursions when English sea power was still relatively weak. But as ever-bigger ships were built and the river tended to silt up, Woolwich downriver was used for the very largest vessels. Chatham and Portsmouth had greater strategic importance during the seventeenth and eighteenth centuries, followed by Sheerness and Plymouth. Deptford saw less investment than the other dockyards, except for a great quadrilateral storehouse which would have dominated the river, were it not for the even more imposing Naval Hospital a mile away at Greenwich. Otherwise the buildings were vernacular and largely wooden. The river was narrow, winding and increasingly overcrowded, so it was difficult to get ships up and down, and to keep them safely offshore. The yard was on a restricted site in an increasingly built-up area, and would be difficult to expand.

But Deptford had certain advantages. It was close to numerous commercial shipyards and able to supervise their naval work. It could survey some of the thousands of merchant ships entering the river every year, with a view to hiring them as naval transports. Its closeness to the London markets allowed

Deptford Dockyard in the 1750s by Milton and Canot. The Grand Storehouse is the square building to the left. The wet dock where the yachts were kept is shown diagonally near the centre, with building slips and dry docks extending from it like fingers. The top panorama shows the masts of the yachts in the wet dock. (Royal Collection Trust / © Her Majesty Queen Elizabeth II 2022, 701582)

it to be used as the main victualling depot for the navy. That also allowed close interest by monarchs, politicians and naval officials in London. Samuel Pepys visited it many times. In July 1662, for example, 'To Deptford, and there surprised the yard, and called them to a muster, and discovered many abuses.'[1] It was used for innovative work throughout the eighteenth century, including the building of the early 74-gun ships and frigates in the 1750s, fitting out most of the ships used on Captain Cook's voyages, and early experiments with copper sheathing. All this made it a suitable base for the royal yachts.

The yachts at Deptford

Its role with the yachts began late in 1660 when Christopher Pett was ordered to build the prototype, the *Katherine*. Other yachts, however, were built elsewhere including the *Anne* and *Charlotte* at Woolwich, the second *Mary* at Chatham and Anthony Deane's *Cleveland*, *Navy* and *Charles* at Portsmouth. The more functional boats *Kitchen* and *Merlin* were built commercially on the Thames. It was only in the next century, largely under the guidance of master shipwright Richard Stacey, that Deptford became the standard building place for royal yachts.

In addition to the yachts for the British royal family, Deptford also built the *Denmark* to be presented to the Prince Royal of that country. In February 1785 the yard was seeking out single and double futtocks suitable for the purpose, and in two months it expected to need plank for the bottom. By August it was nearing completion and Lord Howe, the First Lord of the Admiralty, personally visited the house of Edward Hunt, the master shipwright, to select furniture for the yacht. His lordship was also pleased with a fire hearth being fitted on board. The yacht was ready to launch soon afterwards, 'except some finishing works in the apartments and painting the lower rooms and between decks'.[2] She was returned ignominiously as a protest after the Royal Navy bombarded Copenhagen in 1807, and entered naval service as *Prince Frederick*.

As well as building and fitting the royal yachts, Deptford was responsible for looking after them during the long periods when they were not in use. Policy varied, especially on the question of whether they were to be moored in the river or laid up in the wet dock in winter. Certainly, the overcrowded river could be dangerous. In February 1698 the *William and Mary* was at her moorings off Greenwich when she was hit by the *Harwich* pink causing 'the frame of a light in one of a ?? and part of a cable lost', at a cost of £3. Soon afterwards, she was hit by the *Rachell* of Margate, losing 'two pieces of her fife rail and one of her quarters, some small pieces of frieze and part of her mast cloths'. Nevertheless, it only cost £2 10s.

In 1720 Galfridus Walpole of the *Carolina* reported, 'an India ship broke loose yesterday from Deptford and drove foul us, and broke away our larboard quarter with some other damages'.[3] In March 1737/8 Captain Molloy of the *Mary* wrote, 'This day a laden collier called the *Goodwill* of Yarmouth, Samuel Buffon master, came foul of me on the tide of flood and carried away my bowsprit and did some damage.'

Ice was another hazard. During a cold spell in November 1676 King Charles agreed with 'the commanders of the several yachts now at Greenwich and Deptford, that for the safety of the said yachts they do fall down to Gravesend, for avoiding injuries of the present frost and preventing their being blocked up by it in the River.'[4] In December 1720 the Deptford officers were against wintering in the wet dock, believing that:

> by their lying in the wet dock great damps for want of fire on board will occasion the dews to hang on the ceilings and sides of the cabins, which will very much damnify the Japan, fenering [?] and other works. And if permitted to keep fires on board to prevent the foregoing inconveniency, the wet dock being very full of ships … it may be of dangerous consequences if any accident happen.[5]

On 12 January 1733 Charles Molloy reported, 'This afternoon having a settled hard frost and much ice in the river, I ran the yacht up to Deptford and hauled her into the wet dock and she lay there froze up till the 24the inst.'[6] During the 1720s and 30s Molloy constantly referred to 'my' moorings off Greenwich and only went into the wet dock when necessary.

However, in 1736 Richard Stacey implied it had been normal 'all my time, and I am informed before my coming hither' to put them in the wet dock 'in the winter season to preserve them from the ice' – presumably floating downriver after breaking off from the frozen area above London Bridge.[7] The Milton-Canot print of 1755 shows them in the wet dock, apparently as a matter of course, the only vessels there. But in 1767 Collingwood the master attendant claimed that they had normally lain off Greenwich 'for many years' because 'we have found by experience that in the basin they decay much faster than in the stream, and are also greatly in the way of the works of the yard'.[8] In the following year some shipowners and the younger brethren of Trinity House complained that 'his Majesty's yachts being anchored off Greenwich the navigation of that part of the River Thames is much obstructed'. After consultation it was agreed that 'two yachts may be laid against the upper part of Greenwich, one just below the moorings at Deptford, and three above the moorings that are up and against the moorings which are sunk'.[9] At the end of October 1771 Captain Alexander Hood of the *Katherine* wrote that 'This is the time of year when the yachts are usually received into the wet dock to prevent accidents at the moorings in the winter season'.[10]

The wet dock, much improved since it was first built in

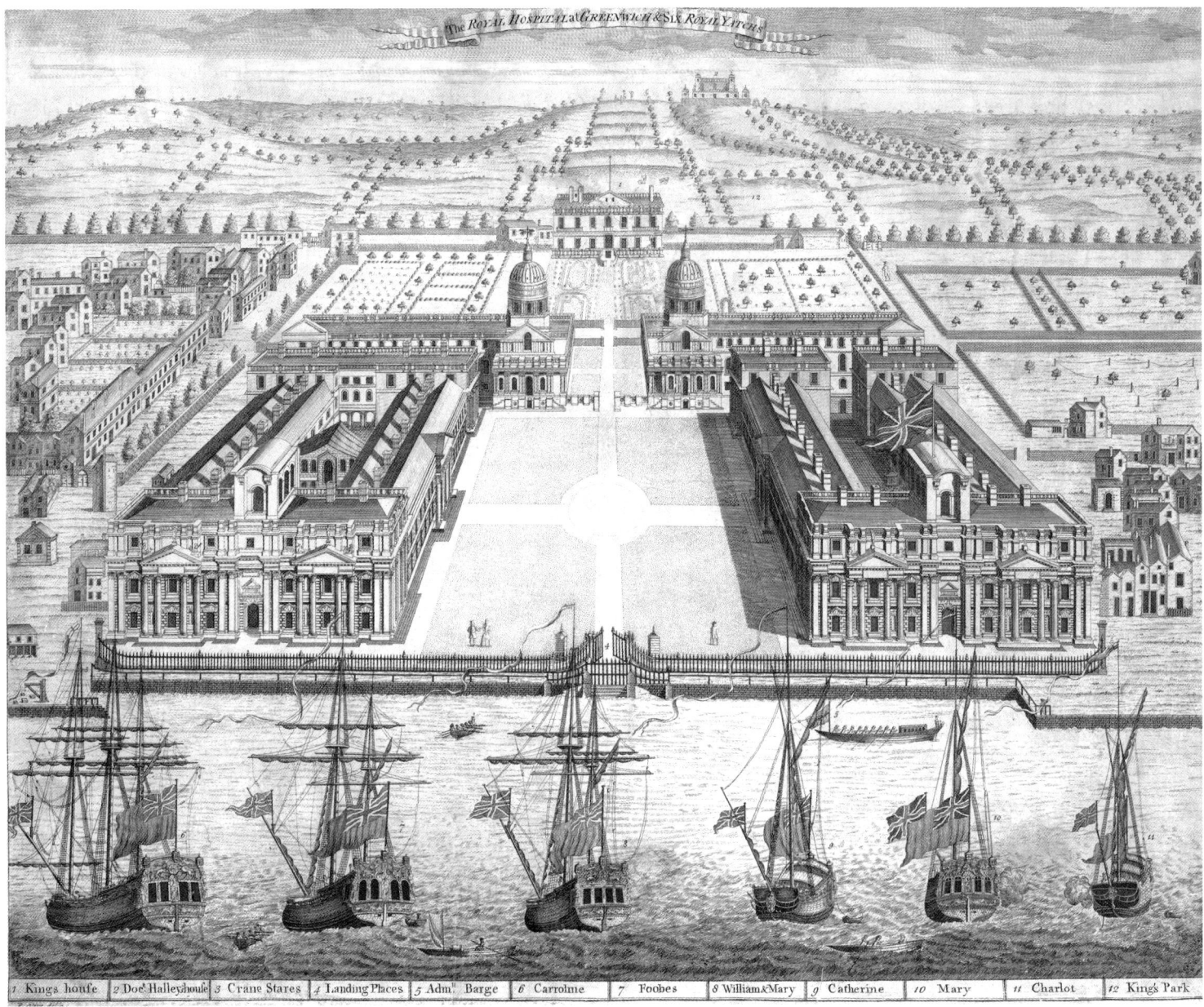

The yachts moored off Greenwich Hospital in the 1720s, at right angles to the bank at slack tide. From left to right they are the *Carolina*, *Fubbs*, *William and Mary*, *Katherine*, *Mary* and *Charlotte*. (© National Maritime Museum, Greenwich, London, PAJ4031)

Henry VIII's time, was hexagonal, rather like a rectangle with two of its corners cut off. It ran roughly northwest to southeast, and its sides were supported by brick faced with timber. It was 280ft long and 215ft wide. It was entered through a gate in the southern corner, which would normally be opened at high tide so that the level inside the dock was kept up. It had three building slips for small ships on its northwest face and it is likely that the yachts were built on them. It also had a dry dock 140ft long and 30ft wide, which was useful for cleaning and repairing the yachts.

∽ Repair ∽

The first stage in a repair was to survey the ship and report what was needed. In 1692 Surveyor of the Navy Edward Dummer and master shipwright Robert Lee described the method of survey for the Navy Board: 'defects are found either by searching all seams, rents, and trenails with a caulking iron, or by boring in the frame with an auger; by observing the ship's cambering or reathing; the pitched seams to spew out it oakum, or by the looseness of rust-eaten bolts.'[11] After that the officers had to consult with the Navy Board to decide the scale of repair needed. With ships in ordinary, Dummer and Lee suggested in 1692 that:

> An ordinary repair is understood to be the annual trimming of the ship in harbour, by caulking all

THE YACHTS AT DEPTFORD AND GREENWICH

those parts which lie to the weather, and laying on of pitch and other mixed stuff of rosin, tallow etc upon the same; and once in three years at the furthest, to dock them and burn off the old matter under water; to search the seams and caulk them as occasion is and to grave them anew, which is to say pay them all over under water with pitch or other mixed matter, with rosin etc.[12]

A more extensive survey, the triennial trimming, was carried out every three years on each ship in ordinary by putting her into dock. This could often lead to further trouble. On 9 May 1733 the *Mary* was in the dry dock to have her bottom searched and cleaned: 'this morning having washed down the bottom the caulkers began to search the same, and finding several planks defective the carpenters took them out and stripped her bottom'; she remained in dock until 24 May.[13]

Beyond the ordinary repair, according to Dummer and Lee, 'An extra repair is taken to be such a defect in a ship's outward matter to the weather, that their frames cannot be preserved nor the ship fit for any service at sea by an ordinary trimming, without stripping such decayed materials of the outside planning and wales'.[14] A ship was usually assessed as needing a 'small', middling' or 'large' repair, though the boundaries between them were not always clear. The *William and Mary* had a great repair at Deptford in 1713/14, costing £2038; a small repair for £365 in 1717; another repair for £1678 on the hull alone in 1724; a great repair in 1736/7 for £2684; a small repair for £886 in 1742 and a further great repair for £3276 in 1747/8.[15] 'Rebuilding' was the ultimate form of repair, though the term was highly ambiguous and changed in meaning over the decades. It would become a purely administrative term, for the replacement of an old ship which had been broken up, often some years earlier. The *Fubbs* was rebuilt at Deptford in 1724 and went from 73ft 6in long to 76ft 9in, and from 148 tons to 157. It is quite possible that no old timber was used in the rebuild.

A dry dock was needed when extensive repairs had to be made to the bottom over several days, when the tide might rise and fall. It was the only way to service the bottom of a large ship, but for a yacht there were other possibilities. One was to take the ship onto a grid just outside the dock gates at Deptford and wait for the tide to fall, as Charles Molloy did with the *Mary* several times, for example in March 1733: 'I hauled the yacht upon the blocks at the dock gates and washed her bottom; in the evening I hauled off again'. Obviously that could only be done for a few hours between the fall and rise of the tide, and for a job that did not involve removing any planks. Another way was to haul the ship up on a building slip, as was done in January 1735.

Tides were needed to raise a ship high enough to put it into a dry or a wet dock, then the dry dock could be emptied as the tide fell, without any need for pumping. Docking and launching were almost invariably done during spring tides, part of the fortnightly cycle. In October 1748 it proved impossible to dock two of the royal yachts – 'this morning tide, through the violence of the wind in the night fell off three feet, so that we could not dock them.' The *William and Mary* was eventually brought in to the stern of the double dock, but 'she touched abaft twenty feet short of the dock gates, and we were obliged to heave her off again.'[16]

Since they did not sail in tropical waters, the yachts had no need for sheathing, an extra layer of thin planking intended to protect against shipworm. Mostly they were graved. That could be done by covering the underwater hull with black stuff based on tar and pitch, white stuff, or a combination of the two known as brown stuff. Yachts usually used white stuff, a mixture of rosin, train oil and sulphur, which was more expensive but produced a more attractive appearance. However, in 1737 the *Royal Caroline* was ordered to be 'payed under water, with stuff tempered with brimstone made of a dark lemon colour'.[17] A yacht's hull might be tallowed on top of that, using animal fat to produce a smooth surface and faster sailing.[18] In April 1732 the log of the *Mary* recorded: 'This morning I hauled the yacht into the great dry dock, washed her bottom and payed the same with tallow.' By the end of the century it was common to copper them, like all the ships of the navy. The *Royal Sovereign*, for example, was coppered soon after her launch in May 1804; it was repaired in 1810 and the following year it was taken off and renewed as part of a middling repair. Part of it was taken off during a refit in 1817, followed by routine repairs in 1818 and 1819, with a further re-coppering in 1820; all that work was done at Deptford.[19]

For fitting the lower masts the yacht was taken into the river and hauled alongside the sheer hulk, defined as: 'An old ship of war, cut down to the gun or lower deck, having a mast fixed in midships, and fitted with an apparatus consisting of sheers, tackles, etc, to heave out to in the lower masts of his majesty's ships'.[20] This was more convenient than using derricks on shore, for the hulk would rise and fall with the tide alongside the ship being fitted.

The *Success* of 546 tons had been built at Deptford as an armed transport in 1709 and was reduced to a hulk in 1731. In 1748 she was sold for scrap, raising £175.[21] Her successor the *Panther* was a 50-gun ship of 715 tons built at Portsmouth in 1716 and paid off at Deptford in 1743. She was converted at a cost of just over £3000. The *Bedford* was a 70-gun ship of 1229 tons and in 1766 she was sailed round from Chatham by ninety men, including dockyard workers and men for ships in the Medway. A mast was also sent from Chatham to equip her.[22] In 1786 it was decided to replace her with the *Worcester*, 64 guns of 1769, which survived off Deptford until 1816.[23]

A typical large repair is described by Charles Molloy of the *Mary* between January and August 1735:

Deptford Dockyard from the River Thames c1775, with the stern of a ship on a building slip, the Grand Storehouse, and the twin domes of Greenwich Hospital in the left background, by Richard Paton. (Royal Collection Trust / © Her Majesty Queen Elizabeth II 2022, 405164)

Thursday 22 January
This afternoon I hauled the yacht out of the dry dock alongside of the hulk, took out my mast and bowsprit, guns, anchors and cables together with officers' stores and hauled into the wet dock …
Friday 23rd
This morning the yacht was hauled up into a slip as per order to have her thoroughly repaired where she lay under the shipwrights' hands until the 30th day of August following …
[30 August]
This afternoon the repairs being finished the yacht was launched. I then hauled her out of the wet dock alongside of the hulk and set her masts and bowsprit …
31
This day I got my shrouds over the mast head, took in my guns, anchors and cables, then hauled into the wet dock again to rig …[24]

The town

Deptford was not a particularly attractive town in the eighteenth century. It had started as two separate villages – Lower Deptford on the river, originally the home of seafarers as well as containing Trinity House, which supervised navigation aids, and Upper Deptford, which serviced passengers and coaches passing over the bridge towards Kent. The two were increasingly joined up during the eighteenth century, Trinity House and most of the seafarers moved away, and the town was more and more dominated by dockyard work. It was a substantial settlement, with a population estimated at 12,000 in 1711 and calculated at 18,000 in 1801, but it had no natural leaders. As Rev James Bate put it rather starkly in 1745: 'We of this parish are partly of middling, but generally of lower life.'[25] The officials of the dockyard were secluded in government houses behind high walls. The occupants of the largest private house, Sayes Court, including the diarist John Evelyn, did not operate as medieval lords of the manor. By 1725 it was let to 'a gentleman who has set up a manufacture, in which we are told he employs near a hundred people'.[26] The town had no local government of its own until 1832, no recognised market place, and none of the guilds which had dominated medieval towns. It had surprisingly good educational and medical facilities by the late eighteenth century, but for many the prosperity brought by the dockyard often did not last into peacetime. Early in the reign of George I, the inhabitants complained, 'Frequent discharges out of His Majesty's dockyard, now that fewer hands are necessary for the public service, leave many artificers and labourers destitute'. The level of poverty in the workhouse can be seen in 1745, when Mary Powland's shift and apron were redeemed from the pawn shop so that she could get a job.[27]

Apart from the modern brick-built Albury Court with houses decorated by wooden carvings, the town was largely built in wood allegedly brought out of the dockyard in the form of 'chips', and the name of 'Tinderbox Row' in Butt Lane gives some idea of its flammability. The town was home to many naval warrant officers, especially the boatswains, carpenters and gunners who stayed with their ships when they were out of service in peacetime. Thomas Lenham, superannuated as gunner of one of the royal yachts, rented a house from Mrs Wickham and was a comparatively good tenant in that 'when he receives his pay, pays his rent.' Naval timber travelled down the main street in wagons and the locals were hard put to maintain it. The dockyard officers had to agree that :

> the road leading from the upper town to the back gate of the yard, which being the only way timber and other heavy goods can be brought by land carriage to the yard, we believe has been thereby much impaired and the putting the same in repair will not only be heavy upon the parishioners, but the want of doing it may obstruct his Majesty's service; some carriages coming to the yard have been broke there last winter …[28]

This was not a route for a sovereign to begin his voyages.

Greenwich

Instead, the King usually embarked at Greenwich. To Daniel Defoe in the 1720s it was:

> the most delightful spot of ground in Great Britain; pleasant by situation those pleasures increased by art, and all made completely agreeable by the accidence of fine buildings, the continual passing of fleets of ships up and down the most beautiful river in Europe; the best air, best prospect, and the best conversation in Europe.[29]

It is doubtful if the rather boorish Hanoverian Kings appreciated this fully, but they must have been aware of the buildings, if only because their governments had to find huge sums of money to build them.

The Palladian Queen's House, avant-garde in its day, was already complete when the first yachts arrived. It was rarely occupied by the eighteenth century, but could serve for royal refreshments or an overnight stay during the journey. Charles had ordered the replacement of the old Placentia Palace beside the river with a much more modern one, which was built in 1662–69. Queen Mary landed there in 1689 and was soon affected by the sight of sailors starving in the streets. She made plans for the Royal Naval Hospital, grandest seaman's hostel in the world, designed by Sir Christopher Wren. Her husband William III was horrified by the expense, but kept it up in her memory after she died in 1694. It was still difficult to find the money – when George I landed there in 1714 it was far from complete. Defoe observed that though unfinished:

> The building is regular, the lower part a strong Doric, the middle part a most beautiful Corinthian, with an attic above all, to complete the height; the front to the water-side is extremely magnificent and graceful; embellished with rich carved work and fine devices, such as will hardly be outdone in this, or any age for beauty or art.[30]

But it was 1751, well into the reign of George II before it was finished.

It has been described by the architectural historian Sir Charles Reilly as 'the most stately procession of buildings we possess.' Arriving by coach, a royal passenger could be carried through the central courtyard between the four magnificent blocks to the stairs at the waterfront, where he would be rowed out to the waiting yacht. Passengers who wanted to avoid the tedious and uncertain passage down the Thames could arrange to meet the yachts at Gravesend downriver, or Margate on the north Kent coast. If heading for northern Europe they could meet them at Harwich, if going south at the Downs or Dover. But they were more likely to disembark at these places, while Greenwich was a favourite spot for embarkation, perhaps because the prevailing southwest wind was more likely to carry the yachts downriver quickly.

3

Rebuild, Repair, Renovation and Replacement

George I and his son left no detailed comments on the yachts they sailed in, and they seem to have been more or less satisfied with what their predecessors left them, with the *Royal Caroline* carrying the King, and the others his ministers and courtiers. The biggest change to the yachts in the early Hanoverian period was in the 'rebuilding' of two of the older ones. It has the advantage of providing detailed plans of both ships, but it is impossible to say exactly how closely they were related to their predecessors, except in dimensions. The term 'rebuild' is always highly ambiguous, but by the 1720s, for warships generally, it was equivalent to ordering a new ship of roughly the same type as the old one, but not necessarily using any of her design or materials.[1] It also overlapped with the concept of 'great repair' or 'large repair', which was executed many times, particularly on the *William and Mary*.

There were six principal yachts by this time, apart from those used by regional governors and the dockyard officers – the old *Isabella* was sold in 1716. Their changing role meant that they were used more in the open waters of the North Sea, and not by sea lovers such as King Charles and his brother, or military men like Marlborough who had learned to endure hardship. In these circumstances, some of the defects of the Stuart yachts became more evident, and the tendency was to increase dimensions when the yachts came in for major repair or rebuild. Ketch rig became more common, and the *Mary* was re-rigged in 1736. There was also a tendency to raise the height between decks, as was done with the *Mary*, *William and Mary* and *Augusta*.

Repairs and rebuilds

The *Fubbs* was surveyed at Deptford in March 1724, and the results were not good: 'The keel, false keel, post and stern decayed, all the plank of the bottom and sides, the strake between the wales on the starboard side, and the upper wales taken off, being decayed.' In addition the frame was 'worn and decayed' though 'some part of the joiners' work, and most of the carvers' work' was serviceable. The dockyard officers recommended that 'it will be for his Majesty's service to take her to pieces and with the serviceable remains to rebuild her'.[2] This was agreed and there were some changes. She was to be fitted out 'in all respects as the *William and Mary*', but in June the Navy Board pointed out that the latter ship, once a principal royal yacht, had a stateroom which was 'adorned with expensive carved works'. When consulted, the Admiralty ordered the Navy Board 'to cause the state room in the *Fubbs* yacht to be fitted with plain joiners' work.'[3] Compared with the original, she had a slightly shorter keel, but was one foot wider and two inches deeper. Her plan is the first scale view of the interior of a yacht. There is also a model which is believed to be of this rebuilt ship.

In October 1726 the *Mary* was taken into the single dock at Deptford and found to be in very bad condition – more than 150 parts of the frame were decayed and '[t]he foremost and after piece of keel, false keel, sternpost, transoms, fashion pieces, stem, false stem, and to breasthooks decayed'. The dockyard officers recommended 'to take her to pieces and with the serviceable remains to rebuild her'. It would cost £3000 for the hull and rigging, and £100 for furniture and stores, and could be completed in four months.[4] In January Deptford sent to the Navy Board 'a draft and a solid shaped exactly according to it by which design to rebuild his Majesty's yacht the *Mary*'. The use of a 'solid' or block model was standard at the time, and it suggests that the new *Mary* was not just a repair of the old. By the 26th of the month they were preparing her frame, which implies that she had not yet been laid down on the building slip. By July they were suggesting completion in September, which was well beyond the four months planned, but might well have been affected by shortage of labour – the officers were soon asking permission for the shipwrights to work 'tides', or overtime, to complete various vessels including the *Mary*. By August she was 'in such forwardness that the joiners work might be gone in hand with, to be in readiness'. But it was March 1728, more than a year after that original proposal, before she was launched into the Thames. Though the height between her decks was already quite generous at 6ft 1in, this was later increased to 6ft 6in.

In November 1743 the Deptford officers had orders to take the *Katherine* into the single dock to repair her, but in June they reported, 'in searching the seams and butts of the wales, plank of the bottom and sides, found the greatest part thereof defective and rotten' – a common enough problem when ships were inspected more thoroughly. They estimated repairs would cost £1620, but as they explored yet further in a 'supplemental survey', they found that another £720 was needed to replace, among other things, twenty-six floor timbers and twenty upper futtocks. The 'footling', footwaling, or internal planking of the lower hull, would have to be taken off to get at the lower futtock heads.

The *William and Mary* was clearly valued, perhaps because her stateroom was 'adorned with expensive carved works'. She had great repairs at Deptford in 1713/14, in 1736/7 and in 1746/7, all costing more than £2000 each, as well as £2700 spent on smaller repairs between 1724 and 1730. She was rebuilt in 1765 and the headroom between her decks raised from 6ft 1in to 6ft 6in.

The *Royal Caroline*

In 1733 it was estimated that repairs to the *Royal Caroline* would cost £4297 during that year alone, the great bulk on workmanship and materials on the hull, but including £560 for furniture and stores. In fact it would cost £7423 5s 3¾d, of which only £3652 7s 14d was spent on the hull.[5] In June 1749 it was proposed to make alterations, and in preparation for that she was surveyed at Woolwich. The officers, including Thomas Slade, who would later design the *Victory*, gave details of her decayed timbers, which included two sections of the keel, the whole of the kelson, thirty-two floor timbers, thirty-eight lower, forty-five second futtocks and fifty top timbers. The planks of her main deck and forecastle would all need replacing, along with many other parts. They concluded that a full repair would cost £1312 for workmanship and £788 for materials.

The Navy Board clearly did not like the bad news, or it had its doubts about the competence of the Woolwich officers in such work, which was usually done at Deptford, and they ordered another survey which would include the master shipwright and his assistant from that yard. It was concluded by the 24th and was slightly more detailed, but no more optimistic. Eight of the floor timbers had been 'shortened and double bored on former repairs' and 'twenty-four more were rotten and decayed', which matched Woolwich's total of thirty-two. Twenty-three second futtocks had been repaired in the same way and the same number were rotten, one more than had been condemned by Woolwich alone. It was similar with the rest of the hull, except the new survey drew attention to the main stern post, which was 'split by the gudgeons, wing transom, the starboard arm being cut formerly by a funnel now thought proper to be shifted'. Repair would cost slightly more, £2200 in total, and clearly the effects of previous work were beginning to show. It was suggested that she might be converted to a sloop which would cost £1493, which the Navy Board pointed out would 'come near to what a new one might be built for', but the Admiralty ordered them to go ahead and renamed her *Peregrine*. At the end of May 1749 Captain Molloy had the harrowing experience of seeing his beloved ship cut to pieces. 'This day the carpenters sawed the ship in two in order to lengthen her 3ft 10½ inches.'[6]

The new *Royal Caroline*

During Lord Sandwich's yard visitation in 1749 it was:

> Resolved that the Navy board do direct the master shipwrights of the several yards each to prepare a draught for building a ship for the reception of his Majesty, taking care he hath proper accommodation, and doth not draw more water than the present ship, and that the Surveyor of the Navy and his assistant each prepare a draught likewise.[7]

Allin, the Surveyor of the Navy, won the competition, perhaps because he was also the leading judge, but his design had many qualities. The new yacht was to be 3ft longer than the old one, and 18in wider, giving a tonnage of 232 instead of 216. The bell-shaped keel section was abandoned but not entirely forgotten; more than half a century later a Mr Kennett proposed 'an addition to ship's keels, more particularly for vessels of an easy draught of water'. The Navy Board rejected the idea, understanding 'that the keels of his Majesty's yachts

A model which is possibly for the rebuild of the *Fubbs* in 1724, showing the slightly more restrained decoration of the Georgian age. (© National Maritime Museum, Greenwich, London, SLR0430)

had formerly some projections of a similar nature attached to them, but which was discontinued on account of the little advantage derived thereof.' Moreover, it was 'subject to inconveniences by injuring the cable when from circumstances not uncommon the ship might ride across it.'

Otherwise the new *Royal Caroline* was similar to her predecessor, with a wide floor angled slightly upwards in midships, a gently curved cross-section, a long, round stem post and very slight hollow in the bows. The design would later be developed by William Bately for the *Richmond* class of frigates in the late 1750s, though his productions were generally considered inferior to those of his colleague Thomas Slade. The new yacht was to be built at Deptford, as was now traditional for such vessels.

The petty officers and men of the old *Royal Caroline* were discharged, and Captain Molloy and the standing officers were ordered to 'inspect the ship now building in the room of the *Royal Caroline*.'[8] She was almost ready for launch in January 1750 and the master shipwright proposed:

> … the joiners works for lodging the officers and seamen being ready to paint, … it will gain time in fitting her for the sea by doing so whilst those works are performing in the wet dock the remaining joiners' works may be completing, which the master joiner tells me will be finished in three weeks' time after launching if the carvers works they are put to be completed …

Thomas Burroughs the master carver was getting assistance by employing all the house carvers he could find in London, plus the ship carvers on the Thames. His bill eventually came to £1100. The ship was duly launched on 29 January and rigged, then sent on sea trials. She returned to Deptford to be visited by the Duke of Cumberland, the King's second son, and Lord Sandwich. But her gilding was still not complete and by 29 March it was hindered by rain. Shipwrights did not usually worry about precipitation, but in this case, 'the gold will blister and tarnish and the blue run over the gold', so it was best to keep her covered up. Her gilding was mostly done by Rosamond Turner and her company, but her demand for an extra £900 was disputed and did much to end the family's relationship with the dockyard. Instead, Mr Stock was paid £267 for completing it.

The *Royal Caroline* proved to be the classic yacht of the age. Her plans were acquired by Frederic Hendrik af Chapman and featured in his great work *Architectura Navalis Mercatoria*.

The *Augusta*

There was little interest in the royal yachts in the forty years after the death of George II, for his grandson George III gave up the visits to Hanover and did not discover the joys of sailing until late in the century. The only major work was an extensive rebuild of the *Charlotte* in 1710, which gained 20 per cent in dimensions and a new name, the *Augusta*. She was launched at Deptford in 1771 but it is not clear who designed her, or how much of the old ship survived. Though 10ft shorter and nearly 50 tons lighter than the *Royal Caroline*, she was second only to her as the principal yacht of the age, being used for Admiralty visits to the royal dockyards and for the transport of royals, including Prince William and the future Queen Caroline. She was rigged as a ketch until 1773, when she had a foremast added to become a full-rigged ship.

After she was used for Sandwich's dockyard visitations in 1771, it was suggested that her decks might be raised without any detriment to her sailing qualities. Deptford Dockyard produced a plan for this at the end of August, with all the decks raised by a few inches and the headroom in the cabins increased from 5ft 10in to 6ft 4in – though this would have been reduced by the depth of the deck beams above.[9] This was approved by the Admiralty on 6 September.[10] But the building of entirely new yachts would have to wait until the next century.

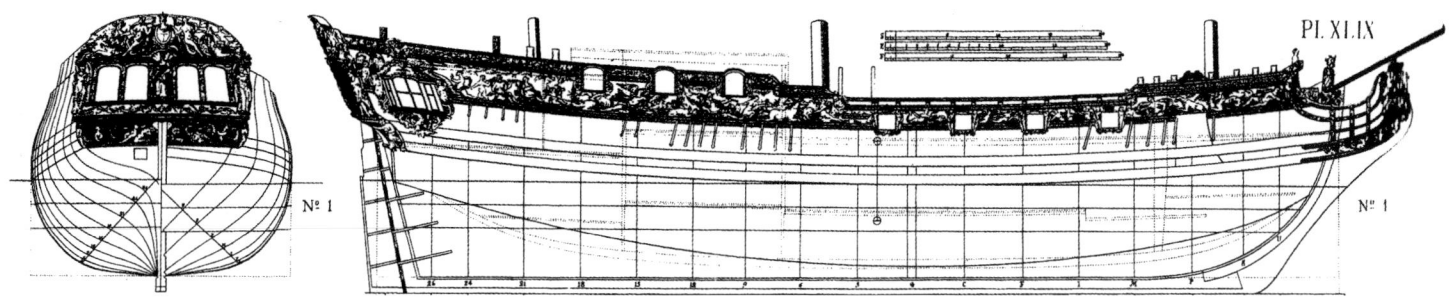

The rebuilt *Royal Caroline* as shown by by Frederic Hendrik af Chapman and featured in his great work *Architectura Navalis Mercatoria*.

4
Other Duties

Luxury transport

When not needed for royal duties, the yachts often carried distinguished passengers, as they had always done. Most were aristocrats, who often held some kind of government or court position – even if they did not, as members of the House of Lords their votes were worth cultivating. There is no sign that members of the House of Commons had the same privileges, unless they also happened to hold high office. It was still very rare for the yachts to travel further than France or Holland, and passengers going further were allocated places on regular warships

The *Mary*'s most regular passenger was the Duke of Richmond, the Master of the Robes in the royal household. In June 1719 the *Mary* moved outside the usual range of ports with orders to pick up the duke and duchess in Chichester Harbour, near their home at Goodwood, and take them to Dieppe. They crossed the harbour bar at eight in the morning of the 9th and anchored just inside. The aristocratic passengers came on board on the 13th and the yacht was towed out part of the way with little or no wind. They repeated the trip in September and landed the duke and duchess by pinnace. They were picked up from Calais early in the following month. The grandees were on their travels again a year later, when the *Mary* was ordered to pick them up from Rotterdam. It took a fortnight to get out of harbour, but after that they had 'a fine gale with smooth water and fair weather', and were soon off Margate. The duke embarked again at Hellevoetsluis in August 1721, and he and the duchess were picked up at Calais in October 1729. The *Mary* received the family baggage at Greenwich on 28 May 1734, then the couple came on board on the following day for another trip to Holland, returning in October. In June 1735 they made the short passage from Dover to Calais in eleven hours amidst 'rain and squalls in the morning and fair weather in the afternoon'. In October 1736 they came on board again at Greenwich and were taken to Rotterdam; they came back in November to be dropped off at Harwich.[1]

Other dignitaries transported by the *Mary* during the 1720s and 1730s included diplomats such as Baron Solenthall, the Danish envoy, from Greenwich to Holland in August 1720; Colonel Dubourgay, the envoy to the King of Prussia, May 1726; and Horatio Walpole and Stephen Poyntz, ambassadors to the court of France, from Calais to Greenwich in September 1730. Walpole, the prime minister's brother and a leading diplomat, also travelled from Margate to Hellevoetsluis in October 1733, escorting the *Fubbs* with the Prince of Orange

Charles Lennox, Second Duke of Richmond, a courtier who often sailed as a passenger in the yachts. (Royal Collection Trust / © Her Majesty Queen Elizabeth II 2022, 657939)

with other yachts; in April 1734 the *Fubbs* picked up the Prince and Princess of Orange at Gravesend, while the *Mary*, *William and Mary* and *Katherine* took their extensive retinue to Rotterdam; the *Mary* returned the next month with the ubiquitous Horatio Walpole. In September 1743 the *Fubbs* was ordered to take Princess Louisa to Holland.[2]

In 1741 Walpole's namesake, the prime minister's son, was returning from his Grand Tour when he found the *William and Mary* at Calais waiting for Lady Cardigan returning from Spa. He was not impressed with her captain, who locked an Italian singer in the cabin 'and swore in broad English that the Viscontina should not stir until she gave him a song'. She refused, and Walpole 'begged her not to judge of all English from this specimen'. Walpole, who was waiting for a wind, apparently travelled in a different yacht whose passengers included 'East India captains' widows' and 'a Catholic girl coming from a convent to be married, with an Irish priest to guard her'.[3] It is unlikely that a royal yacht would have carried such passengers.

Senior military officers were often transported by yacht as they had been in the days of Marlborough. In February 1744 the *Fubbs* was made ready to transport Field Marshal Wade to Holland.[4] In March 1748 the *Katherine* was ordered to take generals the Earl of Ancram, Lord Henry Beauclerk, Lord George Sackville, Lord Howe, Lord John Murray, Colonel Cornwallis and Colonel Conway to Hellevoetsluis. Since their attendants were to travel as well, the yacht must have been very crowded. The *Katherine* was to carry General Sir John Ligonier in February, and the *Fubbs* was ordered to Harwich to pick up the Duke of Cumberland and take him to 'to such place in Holland as he shall direct' under escort from *Hastings* and *Badger*. They were to feed the passengers 'as your yacht's company in their passage' – a standard phrase which did not preclude the dignitaries providing their own cuisine – and to hoist a special flag provided by the Navy Board. The *Katherine* carried General Powlett to Holland in April and the *Mary* took Lieutenant General Hawley in May.[5]

By 1748 both sides had realised that the war had reached stalemate, with the French supreme on land and the British on sea. Preliminaries of peace were signed at Aix-la-Chapelle at the end of April and the navy was issued with orders for 'a cessation of arms between the King and his most Christian Majesty'. The *Fubbs* was sent to get Cumberland back in July, to be followed by Hawley in the *Mary* in November.[6]

Peace and culture

France was now reopened to travel and despite the recent conflict it was still the centre of taste and culture. Paris was a standard visit on the aristocratic Grand Tour, and French was the international language of diplomacy. Even in wartime it was being written that a London mercer 'ought to keep close intelligence with the Fashion-Office at Paris, and supply himself with the newest patterns from that changeable people.'[7]

In June 1749 the *Charlotte* was ordered to take the Earl of Rochford, the envoy to Turin, to Calais and the *Mary* was to take the Earl of Albemarle to the same port on his way to assume his role as ambassador in Paris. Rather than returning to Greenwich as was normal, they were to 'remain with his lordship so long as he thinks fit.' The Duke and Duchess of Richmond were soon on their travels again and the *Katherine* was ordered to take them to Hellevoetsluis, then Calais in August. In April 1751 the *Fubbs* was ordered to take the Duke of Bolton to Calais en route to the south of France to recover his health. And communication was maintained with the Netherlands as a route to northern Europe. In February 1750 the *Mary* was to go to Hellevoetsluis to fetch the Earl of Hyndford and in September 1750 the *Katherine* was ordered to take Lady Margaret Bentinck there and bring the Countess of Hyndford back, though if she did not appear within four days the captain was to put himself under the command of Anson, who was expected soon with the yachts to transport the King. In October the *Fubbs* was to bring back the Duchess of Newcastle.[8] Such activity continued until the middle of the decade, when the threat of a new war made travel far more difficult.

The yachts were still used occasionally for courts martial; in March 1750 the Deptford officers asked to remove 'the contrivances fitted on the *Charlotte* for the admiral to hold a court martial on her'.[9]

Pressing

The first twenty-five years of the Hanoverian era were relatively peaceful, apart from minor wars with Spain beginning in 1718 and 1727. There were, however, several war scares which resulted in short-term mobilisation of the fleet, including the pressing of men. The yachts were useful for this. In peace, normal warships were out of commission for much of the time. The yachts, on the other hand, often remained in commission for many years, though sometimes under different captains. By 1752 the *Charlotte* had been in commission since 1710, the *Fubbs* since 1724, the *William and Mary* since 1730, the *Catherine* since 1745, and the *Mary* and the *Royal Caroline* since 1749.[10] Moreover, they were small enough to go inshore to follow merchantmen and they had experienced and comparatively loyal crews. They were usually at Deptford or Greenwich, where orders from London could reach them very quickly.

War with Spain continued into 1719 and the *Mary* was at Greenwich as usual on 13 June when:

> This afternoon Captain Watts, one of the captains appointed to regulate the press, came on board with a press gang (and orders) to sail down the river to meet the *Cardonnell* a ship from East India, so I loosed from the moorings and run down, and found the ship at anchor in the Hope. I anchored alongside her and assisted rummaging him but found the men (as the information) were all got on shore the evening before.[11]

In August the *Mary*'s captain reported, '7th stood out to sea and spoke with several ships, impressed three men', followed by four men on the 8th; '11th, put them on board the *Endeavour* tender'.[12] A few days later she was off Dungeness when 'three tall ships appeared' – the East Indiamen *Stanhope*, *Grantham* and *Morris*. She followed them into the Downs and then to Margate Roads and the Nore, where she met the *Ipswich* 'from whence I received some assistance of men, so sent the pinnace on board the *Grantham* to impress the seamen

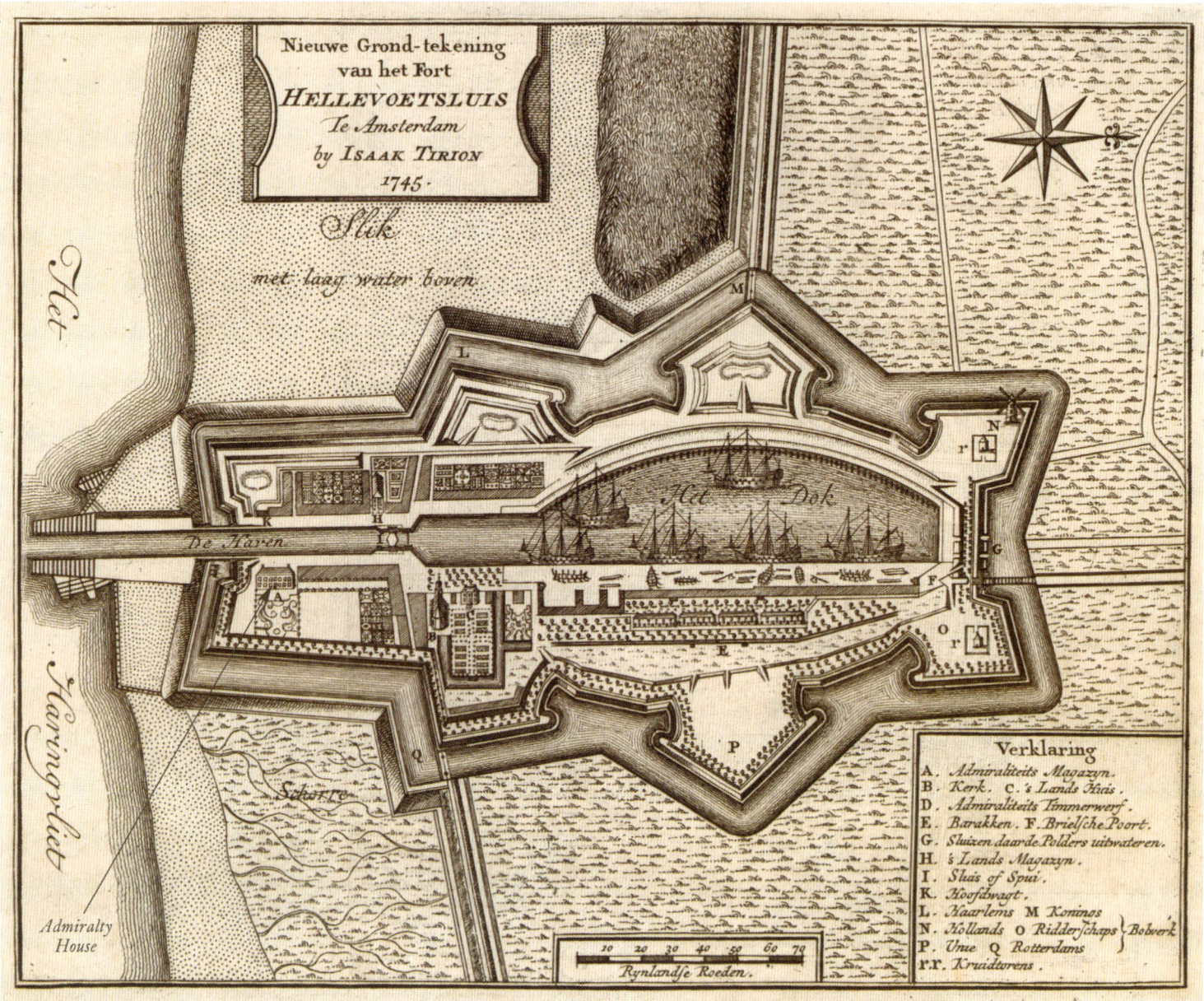

A map of the Dutch port of Hellevoetsluis, dated 1745. The King sometimes lodged in Admiralty House, marked A on the right. (Rijksmuseum)

out of her'.[13] Later at the end of January 1720/1 the *Mary*, '[r]eceived orders to proceed to Gravesend with Lieutenant Brooks to examine the outward bound ships for seamen … at four this morning I loosed from Greenwich and run down, as I anchored at Gravesend, sent my pinnace on board with Lt Brooks to examine the ships there'. Presumably they had some success, for on 7 February they were ordered to go down to the Nore and put the pressed men on board the *St Albans*.[14]

During a later mobilisation, 508 men passed through the *Fubbs* in January to March 1725/6, followed by 163 in April and eighty-four in May and June.[15] And in 1731 the *Mary* was deployed again. On 18 June:

This morning weighed and plied off and on upon the tide and visited several merchant ships. In the evening anchored, when received orders from Sheerness to sail down the Swin after the *Grafton* and take out of her what men she had above her complement and bring them back to the *Cornwall*.
…
At two this morning I weighed and plied down to the *Grafton* and found her under way. I accompanied her the whole ebb as low as the Middle, then took out of her 27 men …[16]

A much more dynamic mobilisation was organised by the Anson Board when war threatened again in 1755. The royal yachts were dragged into this, even more than they had been in the Dutch Wars. In January 1755 the *Royal Caroline* was ordered to send twenty men each to the *Charlotte* and

Katherine, which in turn received orders to go to the anchorage at the Nore and put the men on board the *Culloden*; then to go to the Gunfleet:

> and remain there to raise men, till you have exchanged the aforementioned *Royal Carolina*'s men and as many of your own men as you can spare, in lieu of men you may take from merchant ships bound up the river, when you are to put such raised men on board the *Culloden* and the repair to your usual moorings in the river …

They were issued with press warrants for the purpose and enjoined to employ themselves 'diligently in that service, and to regulate your conduct therein according to such articles of the enclosed instructions as may relate thereto and on no account to procure any men contrary to the tenor thereof.'

In a startling juxtaposition, the yachts were taken off impressment duties to carry the King and his entourage to Holland. They were back by the beginning of May, and fifty men from the *Mary* and *Royal Caroline* were put on board the *Mary* and *Charlotte* which were to go to the Gunfleet to raise men and put them 'on board any of his Majesty's ships at the Nore'.

In the latter stage the yachts were largely superseded by hired tenders, though they still had a role in the process. In October 1755 the *Royal Caroline*, *Fubbs*, *Katherine* and *William and Mary* were to supply twenty to forty men each:

> to be employed navigating from the Downs into the River Thames merchant ships which may have their men pressed from them, and having directed Vice-Admiral Smith to send a tender or cutter up the river to receive so many men from your ship as will complete those you have lent to the Hampton Court, up to the aforesaid number, and also to send one up from time to time as there shall be occasion, to receive and carry back to the Downs such of the said men as shall come up in merchant ships to be safely moored …

The men were to be allowed sixpence per day extra for their trouble.

On 15 November 1756 the *Royal Caroline* was issued with a press warrant and ordered to fit for channel service with an enhanced crew of seventy men and three months' provisions except beer, 'of which she is to have as much as she can conveniently stow'.[17]

Work for the yacht crews

Even in periods of profound peace the yacht crews were not completely idle, for as experienced seamen they were often called to help rig ships in the dockyard or take them downriver, a task which increased in wartime as more ships were commissioned. The master attendant of Deptford Dockyard often had to navigate newly-built or repaired ships to Woolwich or further, using a scratch crew made up of men from the royal yachts, the ships laid up 'in ordinary', dockyard riggers, Greenwich pensioners, and any seamen that could be found. The process intensified in wartime. Typically, on 30 January 1748 Captain Limeburner of the *Fubbs* was ordered:

> to send 25 of your yacht's company under the care of a discreet officer on board his Majesty's ship the *Queenborough* at Deptford to assist in rigging, fitting and sailing her to Galleon's Reach, and from thence to the Nore, taking care to recall them on their arrival there … You are to send the men's clothes and bedding along with them and to direct them to lie constantly on board …

There were similar orders for the men of the *William and Mary* and *Fubbs* at the end of March to work on the *Deptford*, and they were to be paid ninepence per day for their trouble. But the Admiralty felt obliged to inform the captain of the *Deptford*:

> … complaint has been made to us of the negligence of men lent from the yachts, and that many of them having taken a muster do run ashore and absent themselves from duty for the rest of the day, you are in case any of the aforesaid men are guilty of such practices, to secure the most notorious and send them on board the tenders belonging to the regulating captains to be carried down to the Nore among the other pressed men, and you are to let the men know you have orders to do so.[18]

This practice continued after the peace of 1748. Ships ordered during the war in the River Thames had to be completed, and in December 1748 the Admiralty wrote to the captain of the *Katherine*:

> Whereas the Navy Board have represented to us that the works of his Majesty's ships the *Lyme* and *Sphynx* (the former building at Deptford and the latter in Mr Allen's at Rotherhithe) are nearly finished and will be ready to launch on the 10th instant, and that the master attendant at Deptford has informed that there will be a want of 25 good

seamen and an officer that morning in the yard by seven of the clock, to assist in setting their masts and rigging, we do hereby require and direct you to order 25 of your yacht's company, together with a discrete officer, to be in the said yard by seven o'clock tomorrow morning in order to assist in rigging and setting the said ship's masts and transporting them to their proper moorings accordingly.[19]

In 1751 John Goodwin asked for a lieutenant, four midshipmen and 100 seamen from the guardships, twenty men from the *Fubbs* yacht and 120 men from the ordinary at Deptford and Woolwich to sail the 70-gun *Buckingham* from Deptford to Chatham.[20] For moving the *Hector* to Portsmouth in June 1774 the Navy Board proposed to use thirty men from the *Royal Charlotte*, eighty more from the other yachts; nineteen riggers, seventeen 'extra men', forty labourers and a shipwright from the yard; and sixty more men from Chatham and Sheerness. The officers consisted of the master attendant, a master, four boatswains and a surgeon to give a total of 255. The Admiralty agreed that the ship should be escorted by a frigate or sloop 'to give assistance in case of need'.[21]

It was important not to miss the tides for sailing, as in December 1758 'the men belonging to the yachts who were proposed ... to assist in sailing the *Warspite* to Long Reach not having appeared from the *Aquilon* before the tides had fallen off too much for the pilot to take charge of the *Warspite*, was the reason of her not sailing the last spring'[22] – which meant that the ship might be delayed for two weeks before sailing.

Sometimes there were more menial tasks to perform. Ships were laid up after the war and in June the crews of the *Royal Carolina* and *Katherine* were to send forty and twenty men respectively to go 'on board his Majesty's ship the *Deptford* at Deptford to assist in unloading of her stores and putting her into a proper condition to be delivered into the custody of the officers of the yard.'[23] The work was not complete by May 1750, for:

> part of the rigging ... is not yet brought ashore, that the rest with some other stores in the craft at the wharf are not taken out, a good deal of water in her and the *Harwich*, the ships to unmast and ballast to be taken out of them afore and abaft, with several works to be performed to the *Swiftsure* ...

All six yachts in the port were to send twenty men apiece under a boatswain to assist, for a yacht crewman had to be flexible. Sometimes the work was more menial. In January 1748/9 the loading of the *Deptford* storeship (not to be confused with the 60-gun ship mentioned above) was going 'but slowly' and the *Royal Caroline* was ordered to send twenty men to assist. And in April 1751, the captains of six yachts were ordered:

> to send your boatswain, masters' mates, midshipmen o'clock in the morning on the 30th of this month, to assist and as many of the seamen belonging to his Majesty's ship under your command as can be spared, into his Majesty's yard at Deptford by six o'clock in the morning on the 30th of this month, to assist at the launching of his Majesty's ship the *Buckingham*; ordering them to follow the directions of Mr Goodwin, master attendant of that yard, in the performance of that service; and directing your boatswain and masters mates to take care to your ship's men, and also the yachts men who are ordered to attend on the same service, strictly to their duty.

~5~
The Last King in Battle

George II proposed no radical changes on his accession to the throne in 1727. Living in the most conservative part of the century, he offered neither innovation nor strong character. His early attempts to win back some power from his ministers gradually petered out as he aged. He was chiefly famous for being the last British king to be foreign born, and the last to lead his army in battle.

Following the precedent set by his father, George embarked on the *Royal Carolina* at Greenwich, on 17 May 1729. The admiral in command was Lord Torrington, a hero after his stunning victory at Cape Passaro in 1718, but sometimes suspected of Jacobite sympathies. There were five other yachts as usual. They passed along the east coast within the Gunfleet Bank, and sailed close to the wind to reach the Dutch coast by the 19th. He disembarked at Hellevoetsluis on the 20th. On shore, a sophisticated operation was developed to carry the King onwards. He was accompanied by a party of ten or twelve riders and took meals at inns along the road. They hoped to cover nearly two hundred miles to Osnabruck on the first day. Standard instructions were drawn up with blanks for the names of individuals.[1]

Torrington was also in charge of the return trip, as he had told the Duke of Newcastle, Secretary of State, that he 'desired the honour upon this occasion'.[2] The captains needed some skilled seamanship after they reached the Dutch coast: the *Mary* had to veer out two cables to hold her in 'a great sea and stout gale', and the *Charlotte* lost her mast. The *Mary* reached Hellevoetsluis only to be told that the King had already sailed in the *William and Mary*. She sent a pinnace ashore to check but it proved to be true. It was not the last time the King's impatience would cause problems for the yacht commanders.

The King's visits would settle down to a three-year cycle, with the next one in 1732. Though he retained Hellevoetsluis as the principal point on the Continent, he was more flexible than his father about disembarkation on the return journey. In this case, the yachts used the wind to sail along the east coast of England which brought them across the Thames Estuary well to the west of Margate. As a result, on 29 September he was rowed ashore from the Hope to land at Gravesend. 'At ½ past 1 o'clock at the upper end of the Hope, the King went from on board of us in our boat, we struck the standard and saluted His Majesty with 7 guns. Sir Charles Hardy steered the King ashore to Gravesend.'[3]

The ministers were reluctant for the King to go again in 1735. Inevitably it interrupted government business, even though the Queen could reliably be left in charge during his absence. However, it meant that 'every paper, which in that case might be regulated by a short journey only from Sir Robert Walpole's house at Chelsea, to the King's palace at Kensington, being obliged to make a voyage or two from England to Hanover before it could be settled.' Moreover, there was a war in Europe, raising the prospect that the King 'might be running through Westphalia to England with seventy thousand Prussians on his heels.' The King only answered 'Pooh' and 'Stuff' to objections, and he was supported by the Queen, who seemed to relish her time in charge and enjoyed 'being mistress of all those hours that were not employed in writing, to do what she pleased, which was never the case for two hours together when the King was in England.'[4] She did not reckon on the King becoming enamoured with Amelia Sophia de Walmoden, a fashionable married woman in Hanover. Bizarrely, George wrote long letters to the Queen describing his courtship of her. He had despised Hanover while his father was alive, but his affection for his birthplace was growing and now he had an extra reason for going there.

The voyage back in October 1735 was one of the fastest yet. The King left Hanover on Wednesday 22nd and was at Hellevoetsluis at four in the following afternoon. The yachts sailed right away. The voyage across was slowed slightly by the need to anchor during fogs and light winds, but they were soon off Orfordness and entered Harwich Harbour where the King took a coach to London. He was in Kensington on Sunday evening in time to greet the Queen returning from chapel and there was a friendly welcome, despite the King's dalliance. But Lord Hervey, the Lord Privy Seal, believed that 'unreasonably hurrying himself to arrive in England' made him ill. Presumably it was the coach trips that were the main problem, the sea voyage being mostly in calm water.

~ A new mistress ~

The 1736 voyage would prove to be one of the most difficult, in many ways. For one thing it was out of the triennial sequence and it was widely suspected, not without reason, that it was intended to allow the King to spend time with his new mistress, rather than any reason of state. The repeated absence was resented and a satirical notice appeared proclaiming 'that his Hanoverian Majesty designs to visit his British dominions for three months in the spring.' There was even more civil unrest than usual that year, including the legendary Porteus Riots in Edinburgh. There was sympathy for the Queen and another notice appeared seeking 'a man who has left a wife and six children in the parish'. The Jacobite cause was reviving on one hand, while at the same time Frederick, Prince of Wales, was increasingly estranged from his parents but popular with the crowds. The Queen feared that events 'might shake her family's possession of the throne'. The situation was not helped when the King failed to return as usual on his birthday on 9 November.

The preparations for return trip began on 27 November, when Sir Charles Wager hoisted his flag on board the *William and Mary* at Greenwich and headed down river with a typical collection of yachts – *Royal Caroline*, *Fubbs*, *Katherine*, *Charlotte* and *Mary*. They met five men-of-war at the Nore and were at Hellevoetsluis by the 30th. The King was in no hurry to come home, though his romance had proved less satisfactory than he had hoped; he only arrived on the 10th and stayed in the Commandant's House during ten days of westerly winds, with increasing impatience. According to Lord Hervey's second-hand account:

> The King declared if Sir Charles Wager would not sail, his Majesty would go in a packet boat; that he had told Sir Charles *he would go*; and that Sir Charles, in his laconic Spartan style, had told him *he could not*; that the King had said, 'Let it be what weather it will, *I am not afraid*;' and that Sir Charles Wager had replied, '*If you are not, I am*;' that his Majesty had sworn he had rather be twelve hours in a storm than twenty-four more at Hellevoetsluis; upon which Sir Charles told him he need not wish for *twelve*, for *four* would do his business; and that, when the King by the force of importunity had obliged Sir Charles to sail, Sir Charles had told him, 'Well, Sir, you can oblige me to go, but I can make you come back again.'[5]

As a result the King boarded the *Royal Caroline* at ten in the morning of Monday 20th and the yachts and men of war set sail in a south-southeasterly breeze. They passed the small island of Goree and got over the Hinder Bank at the mouth of the estuary at half past one, but at six in the evening the wind veered to the southwest and rose to a 'stiff gale'. Molloy in the *Mary* wore ship at midnight with the other yachts and headed south under reefed mainsail, until that had to be taken in, and he sailed under foresail and mizzen alone in 'hard gales and a great sea'. At ten in the morning, 'it blowing hard the admiral bore away for Goree as also the other yachts except the *Charlotte* who separated in the night'. The King had apparently told Wager that he wished to see a storm at sea and on their return Wager asked if his curiosity was satisfied. The King replied, 'So thoroughly satisfied, that I do not desire to see another.'

Meanwhile, there was increasing anxiety in London. The Queen, who loved her husband despite his wanderings, was distraught. The government was equally fearful for the survival of the regime. On 18 December a report from Harwich mentioned the firing of distress guns off the coast and there was tension until a messenger 'in his dirty boots' arrived to announce that the King had not left Hellevoetsluis. There was more worry on the 20th, the day when Wager was goaded into sailing, when the wind turned from easterly to strong westerly, exactly the phenomenon which had forced the yachts to turn back.

Bad weather continued and the King had no choice but to endure Hellevoetsluis for three more weeks. The point was forced home when a London merchant ship was driven onto the coast, followed by the 40-gun warship *Princess Louisa*, with the loss of sixteen men. The survivors were divided among the other ships for the voyage home, with the *Mary* being allocated fifteen, all with tales to tell. At home, the Queen was in tears on the 26th when Sir Robert Walpole tactlessly caused her to fear the worst, until a messenger arrived with a report of the abortive voyage on the 20th.

At last in the morning of 13 January the weather was deemed suitable and the King boarded the *Royal Caroline*. They made a fast passage, though the *Mary* for one had to sail 'on a bowline', or close to the wind, and they could not head any further south. The King landed at Lowestoft in the morning of 14 January 1737 after an absence of more than seven months.

∽ Wartime voyages ∽

Britain was at war with Spain from October 1739, but their fleet was not regarded as a threat in the North Sea, especially when the yachts were protected by warships. The parliamentary session ended on 29 April 1740, and on the 8th the King's daughter Princess Mary was married by proxy to Prince Frederick of Hesse-Kassel. The King set off for Hanover, embarking at Gravesend on the 13th, but by the 22nd he was still at Sheerness waiting for a wind and summoned the Cabinet to him for a meeting; among the topics discussed were the manning of the fleet for the expanding war with Spain, and the Lords of the Admiralty were ordered to attend as well as the Cabinet.[6] The yachts made a typically noisy departure on the 23rd – 'Vice-Admiral Stuart on board the *Torbay* and the *Cambridge* saluted with 21 guns each, and the fort with 61'.[7] The yachts were convoyed by the *Dragon*, *Gloucester*, *Seahorse*, *Royal Escape* and *Cruizer*, again under the command of Wager. There was the usual fast passage in a favourable wind and the King went ashore at Hellevoetsluis next day.

In mid September the *Fubbs*, *Mary* and *Charlotte* were all in the Great Dock at Deptford being scrubbed and tallowed in preparation for the return voyage. This time they were under the command of Vice Admiral Sir John Balchen and after a slow passage in the Thames Estuary they reached Hellevoetsluis on the 23rd, where they waited for seventeen days. At last His Majesty embarked on the *Royal Caroline* on Monday 11th. They were driven further south and usual and the King landed at Deal on the 13th.

The Battle of Dettingen in 1743, with George II and his son the Duke of Cumberland viewing the scene on horseback on the left. (Royal Collection Trust / © Her Majesty Queen Elizabeth II 2022, 404786)

In 1741 the King boarded at Gravesend on 7 May, but the tide was unfavourable and the winds were very light. The *Royal Caroline* made progress but the *William and Mary* and *Fubbs* went back to Greenwich, where the latter waited for Baron Steinberg, the King's Hanoverian aide. The *Royal Caroline*, *Katherine* and *Charlotte* made a fast passage in both directions under Wager and were back at Greenwich by the 11th. The *Fubbs* sailed later to take Steinberg over. The return passage began when the King boarded at Hellevoetsluis at nine in the morning of Sunday, 18 October. It was dogged by erratic weather, first of all a calm off Goree, then strong winds from the southeast which forced the yachts to reef their sails. The King disembarked at Aldborough the following day, but the yachts had to endure 'a great sea' before taking refuge in Harwich Harbour.

∽ To battle ∽

The first two Georges had always faced resentment about their voyages to Hanover, and they did not have the excuse, as William of Orange did in the first half of his reign, that it was necessary to take part in the war in Europe. That was not the case with George II's 1743 voyage, when Britain fielded a substantial force as part of the 'pragmatic army', an alliance with Austria, the Netherlands and Hanover against France. As well as his dreams of military glory, George was to attend the wedding of his youngest daughter to the Crown Prince of Denmark.

On 25 April 1743 Sir Charles Hardy hoisted his flag on board the *Fubbs* and she sailed from the moorings off Greenwich with the *Royal Caroline*, *Mary*, *Katherine* and *Charlotte*. They anchored off Gravesend that evening and in the following morning the King arrived by coach. He boarded the *Royal Caroline* and knighted Charles Molloy, her captain. They hauled up the anchor and headed out to the Nore to anchor there, but the wind turned contrary in the east-northeast. They took refuge off Sheerness at the mouth of the Medway and after nearly three days, at three in the morning of the 30th, Hardy made the signal to unmoor. They headed to the Nore and down the Swin Channel into the North Sea, where they found a favourable south-southwesterly wind, with 'smooth water and fair weather'. They sighted the coast of Holland in the morning of 1 May and anchored at Hellevoetsluis in the evening. The King went ashore to spend the night in Admiralty House, then set off for Hanover next morning, after his servants and baggage were landed. At four in the afternoon Hardy signalled for the yachts to sail and soon they were off Orfordness with the help of an east-northeasterly wind. They were back at the moorings off Greenwich by six in the evening of the 3rd.

Meanwhile, the King progressed towards the theatre of war, in a convoy consisting of 660 horses, thirteen coaches, thirty-five wagons and fifty-four carts.[8] On 27 June he led his army in the Battle of Dettingen, attacking with sword in hand. He claimed a victory in that the French had lost twice as many men, but it had no strategic effect. George, however, made his mark on history as the last British king to lead his army in battle.

After a typically quiet summer, the *Royal Charlotte* slipped her moorings on 22 October and was hauled into the dry dock to search her bottom in preparation for the next voyage. It was payed with white stuff and on the 27th the crew was paid wages up to 30 June. She was hauled out of dock and stored with four months' provisions. If there was ever any doubt, the purpose was revealed on 2 November when Molloy 'received orders to sail for Holland to bring from hence his Majesty and retinue.' Admiral Hardy came on board and hoisted his flag and they followed the familiar route down the Thames to anchor at the Nore. They sailed in the morning of the 4th with two sloops and the *William and Mary* and *Mary* yachts. They ran over the Flatts and anchored off Margate, heading for Holland in the evening, with 'a fresh gale and fair weather'. The *Carolina* lay by at four next morning 'with the larboard tacks on board', but at daylight they made sail again and anchored off Hellevoetsluis. She moved up to the pier to wait six more days for the King's arrival. They had already moved back into the river, perhaps to make a quick getaway, when he was rowed on board in the morning of the 12th. They put to sea but it fell calm and they went to and fro all night until a 'fine gale of wind' drove them to North Foreland. They soon got as far as Gravesend, where the King disembarked to find that his supposed victory had had a mixed reaction with the public, some of whom suspected him of favouring Hanoverian troops. The yacht went back to her usual moorings.[9]

Rebellion

George Anson returned from his epic circumnavigation in June, very rich from his capture of a Spanish treasure ship. He was knighted, promoted to rear admiral and joined the Board of Admiralty as a junior member, under the civilian Duke of Bedford, and the Earl of Sandwich. He was not even the most senior admiral on the Board, but he was one of the few to have gained any distinction in the war, and he was by far the most influential, which is perhaps why he was chosen to convey the King across the North Sea in May 1745. Like most of his predecessors in command of the royal expeditions, he was not in any sense a courtier: he was famously taciturn, saying little and writing even less, but there was no doubt about his authority and his skill as a seaman.

Britain was now at war with France, and Anson could not ignore the situation; he arranged to convoy twenty merchant ships from Harwich to Holland, though it would prove impossible to keep such a fleet together. From the *Charlotte* he wrote, 'they being all deep loaded vessels, and will not be able to keep up with the yachts, and I cannot delay the King in his passage', he placed some of the escorting sloops to the rear to deal with any French intruders. He sailed on the 10th and was off Hellevoetsluis two days later, ready to land the King in half an hour. He asked the Admiralty for an order to convoy any homecoming ships when he departed.

By the beginning of August the movements of French ships suggested that a Jacobite rebellion was likely, and the Duke of Newcastle told his friend the Duke of Richmond, 'We have repeated our earnest advice to his Majesty to return immediately to his kingdoms, which it is generally thought he will do.' It was increasingly urgent by the 10th when Newcastle wrote:

> I had yesterday three messengers from Hanover, the first of which had been detained some days by contrary winds. He brought orders for the yachts to go immediately to Hellevoetsluis, and directions for getting the Kensington [Palace] ready, for the King's reception. I have hastened the yachts all I can, but notwithstanding I gave a private hint some time ago to the Duke of Bedford [First Lord of the Admiralty], I hear they will not be ready before Wednesday next, and they are at a loss to get a proper convoy.[10]

Three days later he received news that Prince Charles, the Young Pretender to his enemies, had landed in the west of Scotland. In the middle of the month Anson ordered the ships of his squadron to meet him off Hellevoetsluis after completing convoy duties, and the King was back by 31 August.[11] He faced the most dangerous period for his regime, until the rebellion was finally defeated at Culloden in 1746.

Peace again

With the prospect of peace in 1748, the King was already looking forward to resuming his trips to Hanover. Following in the tradition of appointing the most distinguished sailor of the day to command the expedition, in April 1748 George, now Lord Anson, was ordered to take four ships and the yachts *Royal Caroline*, *William and Mary*, *Katherine*, *Charlotte*, *Fubbs*, *Mary* at the Nore under his command:

> … hoisting your flag on board such one of them as you shall think fit. And when his Majesty is embarked, you are to convoy him over to such port in Holland as it shall be his pleasure to land at, taking under your care and protection all such ships and vessels as have been take up for the service of his Majesty's household, or for the passage of such persons of quality as shall go over in company with him.
>
> When the *Royal Carolina* and yachts have landed his Majesty and his attendants with their servants and baggage you are to direct their commanders to

A Dutch chart of the Maas from 1666, showing Goree Island and 'Hellevoet Sluys' with its harbour and anchorage to the east.

return to their usual moorings in the River of Thames, coming over yourself in a man of war or yacht as you shall think fit to the Nore.

After which he was to strike his flag and come on shore.[12]

The orders would not have come as a surprise to Anson as he was a member of the Board which drew them up. On 2 May Anson ordered the *Hampshire* and *Hastings* to proceed to the Gunfleet anchorage off the coast of Essex, and the smaller *Greyhound* and *Badger* 'to lay constantly upon the warp and to weigh the moment the yachts come in sight'. He hoisted his flag in the *William and Mary*. But the wind settled in the east and by the 13th the small fleet had to put into Harwich to await a change, though there was no prospect of this by the 18th. At last on the 19th it turned southerly and Anson decided to sail with the noon tide. They had a 'tedious passage and arrived and did not arrive at Hellevoetsluis until five in the morning of the 22nd'. The ships dispersed to their separate stations, and the yachts back to the Thames.[13]

On 11 October Anson received further orders.

The King's pleasure having been signified to us by his Grace the Duke of Bedford, that the yachts with a proper convoy should be at Hellevoetsluis by the first of next month, in order to bring his Majesty over to England, you are hereby required and directed, in pursuance thereof, to take under your command [for ships, two sloops and the *Royal Carolina*, *Fubbs*, *Mary*, *William and Mary*, *Katherine* and *Charlotte*] whose captains are directed to obey your orders and repair with them to Hellevoetsluis so as to be there by the first of next month, and to wait there for his Majesty's arrival.

When his Majesty shall be embarked, you are to convoy him to such port of England as he shall direct, and having so done, you are to come with the yachts and such of the ships as you shall judge proper to the Nore, sending the others to the Downs or to Spithead.[14]

By the 27th he was getting under weigh from Gravesend and he arrived at Hellevoetsluis on the 29th to meet the deadline. However, it was the 21st before the King arrived and he was embarked. There was a fast passage with a northeasterly wind and within twenty-eight hours he was landed near Margate.

~ The hazards and pleasures of the sea ~

On 30 March 1750 Anson received his orders, drawn up as usual by the Board of Admiralty and signed by his colleagues Sandwich, Villiers and Trentham. This time he was to take His Majesty 'to such port in Holland as it shall be his pleasure to land at.' Again there were four ships, two sloops and six yachts, whose captains had orders to embark 'such persons, foods and necessaries as his Grace the Duke of Grafton, Lord Chamberlain, shall direct.' On 6 April Anson, still at the Admiralty, ordered 'the yachts to proceed from the river the 11th instant, and the men of war to proceed from the Nore on Monday next, the former into Hollesley Bay and the latter into the Rolling Grounds in order to join his Majesty when he gets out of Harwich harbour.' He hoisted his flag on board the *William and Mary* again on the 15th at Harwich and awaited the King, who embarked in the afternoon of the 16th.[15] They sailed but were obliged to anchor in Hoseley Bay, then they plied off Orfordness before sailing in a northeast wind, which was enough to carry them over towards the east – 'stretched over with the larboard tacks on board and reefed topsails all night, had a short tumbling sea and fair weather.' It was a fast if uncomfortable passage, and they landed the King next morning.[16]

The voyage out to bring the King back proved to be very difficult. Anson hoisted his flag on the *William and Mary* at Gravesend on 5 October, arrived at the Nore next day with the yachts, to be joined by the *Assistance*, *Amazon* and three sloops. They were unable to sail the next morning due to 'the wind blowing extremely hardy at ESE, and it rather increased in the afternoon with the appearance of dirty weather and a great sea.' Anson had faced far worse weather in rounding Cape Horn in 1741, but he thought it unsafe to trust the fragile yachts in these conditions and he took them into Sheerness Harbour as Hardy had done seven years earlier. The *William and Mary* was 'weakly handed' as well as having a boatswain's mate and three men sick, while the *Charlotte* had 'three men sick also'. Furthermore, the *Katherine* 'made a considerable quantity of water yesterday at the Nore' and there were fears of some damage in her bilge 'by falling down when ashore at Deptford.' Anson ordered Sheerness Dockyard to examine her fully and it was found that 'the garboard strake on the larboard side was open, which was probably owing to her fall, and that the oakum in some other of the seams were slack, with a small knot over a treenail on the starboard side open, which seems to have been an oversight in the officers of Deptford when she was cleaned there'.

By the morning of the 11th the wind was coming round and the weather seemed about to improve, so Anson hoped to get to the Nore and then 'as low as Harwich'. The winds remained unhelpful and he only arrived at Hellevoetsluis on the 17th after losing contact with a ship, two sloops and the *Katherine* yacht. The King arrived at Hellevoetsluis on 2 November and stayed in Admiralty House to await a favourable wind. He embarked on the 3rd and they sailed at five in the afternoon. Again they sailed close to the wind with tacks on board but had a fast passage to Harwich, where the King landed.[17]

On 17 March 1752 Anson, now First Lord of the Admiralty, was issued with the customary orders to convoy the King to Holland.[18] He gave sailing orders for his squadron. He would hoist his flag in the *William and Mary*, with the *Royal Caroline* carrying the King half a mile astern. The *Swift* and *Grampus* sloops were to be half a mile ahead of the admiral, on each wing, with the *Vulture* sloop a mile ahead. The *Seahorse* and *Surprise* were to be abreast of the King on either side, and in case any strange ship should attempt to speak to His Majesty 'they are to prevent it and to bring the stranger down to the admiral'. The *Torrington* was to 'bring up the rear and to take care of such vessels as may be hired for the service of his Majesty's household.' All ships were to show distinctive lights at night, and '[i]n case of a separation by any unavoidable accident the place of rendezvous is Hellevoetsluis.'[19]

It proved to be a great contrast to the 1750 trip. On 7 April he landed the King at Hellevoetsluis 'in perfect health … after a very pleasant passage.' The return passage was equally successful and took less than twenty-seven hours. At half past two in the afternoon of 18 November he wrote, 'I had the honour to embark his Majesty at Hellevoetsluis yesterday at ten o'clock in the morning, and that I have this moment landed him at Gravesend in perfect health.'[20]

The last voyage

In 1755, as another war loomed, the wisdom of the King leaving the country was debated yet again:

> The French armaments, the defenceless state of the kingdom, the doubtful faith of the King of Prussia, and, above all, the age of the King, and the youth of his heir at so critical a juncture, every thing pleaded against so rash a journey. But, as his Majesty was never despotic but in the single point of leaving his kingdom, no arguments or representations had any weight with him.[21]

In a startling juxtaposition, the yachts were taken off impressment duties to carry the King and his entourage. On 28 April 1755 he embarked at Harwich and Anson set sail. They were at Hellevoetsluis next day, with the King 'in perfect health' as usual. The yachts were back in home waters by the beginning of May, and pressing again. In view of the situation, the King had to come back earlier than usual despite telling one of his ministers, 'there are kings enough in England. I am nothing there, as I am old and want rest and should only go to be plagued and teased there about that D–d House of Commons.'[22] Again the return voyage was pleasant enough, Anson wrote on 16 September, 'I had the honour to embark the King at Hellevoetsluis about nine o'clock yesterday morning, and have the pleasure to acquaint you … that I am this moment stepping into my boat to attend His Majesty on shore at Margate.' He was clearly becoming anxious about being away from London during a growing crisis – soon after landing he wrote to the Admiralty, asking them 'to give me leave to go to town to attend the business of the Board'. George II did not know it yet, but he had made his last trip to Hanover, which would see no more of reigning British monarchs for many years, and the function of the royal yachts would change again.

Part IV

Marriage, Business and Leisure

1

A New Queen

Having ascended to the throne on the sudden death of his grandfather in 1760, King George III had to find a queen 'to procure the welfare and happiness of my people, and to render the same stable and permanent to posterity'. As usual he looked among the royalty of the numerous Protestant German states. He settled on Princess Charlotte of Mecklenburg-Strelitz who, he told the Privy Council, was 'a princess distinguished by every eminent virtue, and amiable endowment, whose illustrious line has constantly shewn the firmest zeal for the Protestant religion, and a particular attachment to my family.' Though George had never met her, a newspaper report described the 17-year-old as:

> a princess greatly esteemed … for unfeigned piety and exemplary humanity and benevolence; her person graceful and handsome, between two extremes of tallness and its contrary; of a very pleasing mien, grace and grand deportment, intermixed with much unaffected ability and composure of mind:– a proper match for so graceful, good and great a monarch as the British King, George III.[1]

Preparing the yacht

In July 1761 the Navy Board ordered that the *Royal Caroline* be 'fitted for the sea with all expedition and for that purpose to make use of all extraordinary means' on her, as well as the *Katherine* and *Fubbs*. King George was a severe critic of the ongoing Seven Years War, which he privately regarded as 'bloody and unnecessary', as well as expensive. This did not prevent him lavishing money on the expedition, and especially on the yacht *Royal Caroline*. It was renamed *Royal Charlotte*, which inspired verses in the press.

> See by what easy deaths great names expire!
> How new ones rise to gratify desire!
> Such titles are, and all the blaze of fame;
> They thus all banish with the loss of name.
> Of kingdom sunk, no traces now appear
> To mark, existing, what their splendours were.
> What matter, then, if we are great or small,
> Since time and chance make nothing of us all.[2]

Royal craftsmen were employed to furnish the yacht, rather than those normally hired by Deptford Dockyard. Samuel Norman of Soho Square, the 'sculptor and carver to the King', was to carry out the perpetual task of regilding. His work was described in three and a half pages of closely written text, which included gold leaf in the stateroom, its companionway and the staircase and passage to it, the awning room, the 'after room or state room', the inside of the quarter light, the forecastle, head, sides and stern. Extra work included:

> … the fore, main and mizzen channels and shrouds, and for the fore and main backstay stools and backstays, blocks rack, fore jack, fore and main topsail halyards and cat: ends of the bumpkins, accommodation ladder, watch bell and three spare trucks for the flag staves at the mast head.

Norman had a reputation for overcharging his numerous aristocratic clients, but the officials of Deptford Dockyard asserted his bill of £1200 was the lowest price he could afford.[3]

The royal upholsters, Vile and Cobb, were 'making and quilting a white satin casting quilt for the Queen's bed in a very pretty figure.' Catherine Naish, the 'royal joyner and chair maker', was to produce a 'settee bed', cushioned and curtained with crimson damask, presumably as a day bed for the distinguished passenger. According to the newspapers, the yacht was to have 'a most sumptuous bed of velvet, laced with gold lace'. In another report, the royal apartment was sealed up and Anson would deliver the key to the princess when she came on board. Horace Walpole told his correspondent Horace Mann that the refurbishing would 'turn the Queen's stomach' – presumably indicating his disapproval of the taste.

The fitting of the yacht attracted a good deal of attention and it was reported that householders around Deptford Dockyard were letting out rooms to see it, 'which very few could'. That is not surprising in view of the high wall round the yard. Sailors usually wore their own clothes or 'slops' bought from the ship's purser, but this time the crew was issued with uniforms 'with gold laced hats, button and loop. Light grey worsted stockings, buckles and pumps, at his Majesty's expense'. The King planned an elaborate entry to the kingdom for his future queen. She was to disembark at Gravesend and progress up the River Thames in a highly decorated royal barge with a band playing and be received by His Majesty in the Queen's House at Greenwich.

The Earl of Harcourt, once George's incompetent tutor, was chosen as ambassador to Mecklenburg-Strelitz, to negotiate the marriage agreement and fetch the princess across the North Sea. He set off early in the yacht *Augusta* and was in the River Elbe by 6 August, bearing gifts including the King's own picture, 'richly and most prettily set round with diamonds, and a diamond rose'.[4] Lord Anson, as usual, was to head the maritime aspect of the expedition. The party was to include Sir Francis Drake representing the Board of Green Cloth which was to cover her expenses, and several ladies-in-waiting. According to Mrs Isabella Ramsden:

> I suppose you know the two duchesses appointed to convoy her intended Majesty. That of Hamilton has made great interest to take her ass with her, which she has done and it is hoped may come back alive. The Duchess of Ancaster only takes a surgeon and midwife, as she is breeding and subject to hysteric fits. They are not to go on shore nor come back in the yacht with the Queen, so will only pay their compliments at her going on board her own yacht.[5]

In fact, the Duchess of Ancaster's sixth child was born just after the voyage on 7 August and named Georgina Charlotte, a compliment to the royal family.

The voyage out

The yachts sailed from Greenwich on 4 July and Hugh Hughes, recently appointed to the obscure office of Coroner of the Verge, was feeling 'very queer but not sick', even in the sheltered waters of the Thames. The fleet assembled off the port of Harwich, the yachts inside the harbour itself, the smaller warships in the Rolling Rounds just outside, and the *Nottingham* and *Winchester* of 60 and 50 guns respectively in Hollesley Bay along the coast to the northeast.

Meanwhile, Anson left London on 6 August, carrying detailed instructions from the King, consisting of 'a full sheet, all writ with his own hand', for his Majesty was 'in haste for his new queen'. Anson boarded the *Royal Charlotte* at 1.45 in the afternoon of the 7th and wrote to the Admiralty, 'I am now getting under sail with the yachts and cutters'. He picked up the other vessels on the way and they sailed in formation with the smaller warships *Hazard*, *Lynx* and *Tartar* in the van followed by the *Royal Charlotte*, the two-deckers *Winchester* and *Nottingham* in the centre, followed by the other yachts, and the *Minerva* frigate bringing up the rear. The country was still at war; the French fleet had been heavily defeated at Quiberon Bay in 1759, but a raid was still possible. On Sunday 9th Anson ordered the *Winchester* to investigate sails to the northeast, which turned out to be a British convoy from the Baltic. The *Hazard* lost some of her rigging in strong winds, and had to be towed by the *Winchester*. Hugh Hughes was recovering from seasickness on the 10th – 'got up was somewhat better eat but little'. He witnessed a near collision between the *Fubbs* and the *Katherine*. On the 13th the seaman on the mainmast sighted Heligoland, or Holy Island as some of the party called it. Pilots came out to meet them, for the entrance to the Elbe was tricky, and the *Fubbs*'s one was fed with bread and cheese. Hughes was pleased to see a porpoise for the first time, but was still longing to get into harbour. He fortified himself with Madeira, which he believed cured his *mal de mer*, but lamented that 'our bread is very mouldy, quite out of small beer'. Some of the cooks were sick, but they were better off than the *Katherine*, which had been on short allowance for three days. There was another accident when a foot-boy fell overboard from the *Charlotte* and was presumed drowned.

Hughes was impressed with the River Elbe, which he believed was 'a fine one, it is much larger than the Thames'. They followed the narrow channel under the guidance of the pilot, though he spoke no English. They sounded the depth constantly with the lead to avoid the steeply rising Vogel Sand to the north, and reached the Red Buoy which marked the shallowing of the channel. The larger warships anchored there; the yachts and sloops proceeded up to Stade, which was already familiar to some from previous voyages. They anchored there on the 15th and a messenger was sent to inform the princess. Hughes went on shore: the town did not answer his expectations, though later he had 'very good bacon and eggs'.

On the 22nd the future queen arrived at Stade and 'there was not a house in the town but was lighted.' She rested for a day, then was rowed down the narrow channel leading from Stade to the Elbe. An observer on board one of the yachts noted:

> At half past ten came in sight the Admiralty Barge, with the Royal standard of England flying in the bow, preceded by Lord Anson's barge with the union flag in her bow. The *Royal Charlotte* yacht was dressed up in all the different colours of all nations to receive her; and the moment she came on board they were down in an instant, and the royal standard was hoisted at the mainmast.

The yachts assembled at Stade to pick up Princess Charlotte. The two single-masted vessels to the left are the *Katherine* and *Fubbs*. The Royal Charlotte is in the centre, bedecked with flags. The *Mary* and *Augusta* are to the right. The princess is in the boat flying the Royal Standard heading toward the *Charlotte*, presumably under the awning. The *Charlotte* is ready to hoist the Royal Standard on her arrival. (Royal Collection Trust / © Her Majesty Queen Elizabeth II 2022, 750449)

She presumably went up the accommodation ladder so expensively gilded by Samuel Norman.

According to the log of the *Augusta*, which showed the common discrepancy about timing: 'At 10 AM Her Serene Highness the Princess Charlotte of M S embarked on board the *Royal Charlotte* yacht, the yacht being full dressed with colours. … We saluted HSH with 21 guns, a general salute was given throughout the squadron and the union flag was hoisted on board the *Lynx*.'[6] To the observer it was 'the finest sight I ever saw'. The princess commented modestly, 'Can I be worthy of all these honours!' which moved the Duchess of Ancaster to tears. If Anson truly had the seals to the cabin, this was the moment to break them and show the princess her luxurious apartment.

They sailed at nine next morning, the 25th, past Glückstadt which did not salute them, and anchored overnight to await the tide. Ritzbüttel and Cuxhaven did salute, and the princess's younger brother was put ashore at the latter port. They hoped to join the larger ships at the Red Buoy but the wind was in their teeth and they anchored, to ride out a gale which lasted through the next day. Hughes was shocked when a man was found dead in a cabin and was hastily buried ashore at Cuxhaven. They sailed again at five in the morning of the 28th to join the larger ships five hours later; Anson transferred his flag to the *Nottingham* for the voyage.

They were not far out when they hit even worse weather – it '[b]lew a storm NW all night, and the greater part of next day.' Hughes recounted 'being awake in the morning with the dreadful sound of a hurricane; indeed it was dreadful for we thought we should be lost, for the oldest man on board hardly saw any such thing.' Many of the ships separated, and Hughes commented of the rest, 'most of the ships in the fleet had some of their rigging or part of their mast lost, the water came over our deck and into the state room. I was very ill indeed and so was most on board'. Nevertheless, 'The *Royal Charlotte* outsailed us all.' Anson as usual said nothing about the effects of the seas. Twenty years ago he had spent a month trying to round Cape Horn in far worse conditions, and in a ship quite similar to the *Nottingham*. In the yacht, art and luxury did

A NEW QUEEN

The storm during Charlotte's passage to England, probably the first painting by the Liverpool artist Richard Wright. The newly-renamed *Royal Charlotte* is in the centre flying a standard. (Royal Collection Trust / © Her Majesty Queen Elizabeth II 2022, 405525)

nothing to compensate for stormy weather: the duchesses of Ancaster and Hamilton were said to be 'very much out of order'. But the princess, according to one report, 'was extremely well the whole voyage' and 'bore the sea like a true British Queen'. According to Harcourt's son, who was on board the *Lyme* at the time, she was 'gay and lively playing on her harpsichord and singing and laughing all day long.'[7]

There was yet more tragedy on the 2nd as a man on the *Nottingham* fell overboard. The weather got worse again, a 'perfect hurricane' according to Hughes. He complained that they were on short allowance again and forced to eat sea biscuit instead of bread. Meanwhile, the King fretted about the lack of news, constantly raising a handkerchief on the end of his whip to check the wind direction. Flamborough Head was sighted by the lookouts of the *Hazard* on the 3rd, but there was another squall. At last a coastguard at Lowestoft sighted the fleet, noted its royal flags and sent a message to the King. Anson decided to take his ships into Harwich to avoid any more hardships and wrote to the Secretary of the Admiralty in his own rather spidery hand: 'You will confirm their Lordships of my being arrived in the port of Harwich, the wind continuing contrary and the Princess being much fatigued, made it absolutely necessary to land her Royal Highness here, where I expect his Majesty's commands tomorrow for that purpose.'[8]

The King's plans for a procession up the Thames were ruined, but he did his best to salvage the situation. Charlotte progressed through the towns of Essex to an ecstatic welcome and married the King on 8 September, six hours after her arrival in London. By Hanoverian standards it was a reasonably happy marriage, producing fifteen children. But many years later Queen Charlotte still had a dislike of the sea.

~2~
Sandwich's Visitations

In June 1749, after the dust had begun to settle after nine years of war, the Lords of the Admiralty, particularly the First Lord, John Montagu, the 4th Earl of Sandwich, were concerned about:

> … the number of men borne in the several dock and rope yards, the great expense attending to the same, and that the works are not carried on with the expedition that might be expected from them, which must arise from the remissness of the officers, or insufficiency of the workmen, or both.

They resolved on a tour of inspection and were rowed to Woolwich on the 26th, where they 'walked round the yard to observe the workmen, many of whom were idle'. They noticed the poor condition of the *Royal Caroline* yacht and ordered her replacement before going upriver to Deptford on the 29th. After a thorough inspection, on 4 July:

> Came on board the *William and Mary* yacht at Greenwich at three o'clock, were saluted with seven guns, the yacht wearing a common pendant, got under sail immediately. Arrived at the Nore at half past seven, the *Surprise* riding there. Captain Baird came on board the yacht, delivered his weekly account and saluted with 19 guns, the yacht returned seven. Arrived at Sheerness at ¼ past nine, and took in the moorings, the officers of the yard came on board to receive the Lords' commands, and were acquainted they would be on shore early in the morning

There was another inspection and at seven in the evening of the 5th they boarded the yacht again. The moorings were slipped and they worked up to Gillingham where they anchored around midnight. By 9.45 on the 6th they were 'against the dock at Chatham'. However, Sandwich's health was causing concern and he returned to London in the yacht, while the other lords went ashore under the leadership of Lord Anson. And on the 8th '[t]hey adjourned the board to the office in London, for which place the Lords went directly'.

Sandwich's health having recovered, they arrived by coach at Portsmouth on 1 August and were saluted by the garrison. They inspected the Royal Naval College which trained young officers, and they had noticed the *Portsmouth*, the old dockyard yacht of 1702. 'The young gentlemen petitioned the Lords that as the old yacht lay in the harbour useless, she might be rigged snow fashion with small masts and yards such as they could rig and unrig upon occasion.' The Lords agreed on condition that there were 'careful persons to attend to see they receive no damage for want of skill.' Their Lordships set out for Plymouth by coach on the 5th, arrived on the 9th and remained there until the 13th, when they set out back for London.[1]

There was no visitation in 1750, but in the morning of 22 May Sandwich boarded the *Royal Caroline* at Deptford with a distinguished party, which included the King's brother, the Duke of Cumberland, Princess Amelia, the Duke and Duchess of Bedford, the Duke of Grafton, and 'several other ladies of quality'. They sailed downriver against contrary winds, anchored when the tide turned against them, and returned to Deptford by eleven that night to put the dignitaries into their barges to be rowed back to London. Sandwich left the Admiralty in June 1751 and his successor Lord Anson had less interest in the dockyards.

John Montagu, 4th Earl of Sandwich, as First Lord of the Admiralty. He is shown in the Royal Naval Hospital at Greenwich, from where he often set off for his dockyard visitations. (© National Maritime Museum, Greenwich, London, BHC3009)

~ 1771 ~

When Sandwich returned to office as First Lord of the Admiralty in January 1771, he was as determined as ever to find out what was going on in the dockyards and to institute reforms, so he soon resumed the series of annual visits which are well recorded in minute books and letters. They used the yachts extensively, though Sandwich gives little detail of the actual voyages. He was to travel in the *Augusta*, newly rebuilt with a considerable increase in dimensions, and ship rigged.

On 10 May Lieutenant George Stoney of the *Wells* cutter, which was ordered to attend the *Augusta* on the tour, asked to take on pilots and fishermen, 'needed to enable him to execute any service required.'[2] The trip in the *Augusta* began at Chatham three days later and the party included Lord Palmerston and several members of the Navy Board. They sailed to Sheerness, and enquired into the depth of water in Medway. On 18 May, according to Captain Bickerton, 'The *Juno* merchantman ran on board us when we lay at anchor off Margate and carried away our bowsprit and head'. Her captain was 'extremely insolent', though he was clearly responsible for the accident.[3]

They arrived at Spithead on 21 May and moored alongside the hulk at Portsmouth the following day, and landed to a guard of marines. There was a five-hour muster of the men, but the yard was seen to be in great disorder due to recent fire. On 28 May they sailed to Plymouth and arrived on 5 June, with the *Cruiser* sloop, *Grace* cutter and *Fly* sloop.[4] Viewing the effects of a large-scale 'plan of improvement' of the yard, they were impressed by a 'very ingenious model of the whole carved in wood by the foreman of the yard'. Sandwich noted, 'it would be very proper to have a plan of each of the yards taken in the same manner'. The original idea was to put them in the Navy Office, but soon it was decided to present them to the King. As a typical Hanoverian he was always more interested in the army than the navy, and was more likely to criticise the design of a uniform button than the characteristics of a ship of the line. Sandwich would soon resolve to change that.

The yacht arrived back at Deptford on 17 June and took up its mooring. The Lords of the Admiralty got into the ship's boat and were rowed up to Westminster. Sandwich's party returned to Deptford in the afternoon of Saturday 22nd, to find five yachts in commission plus the *Royal Charlotte* under repair on a slip. The party did not visit them but were taken downriver to Woolwich to view the dockyard there. They sailed back to Deptford, then, 'At 7 this evening returned to town, having finished the visitation of all the yards upon which we set out exactly six weeks ago.'[5]

~ 1772 ~

Preparations for the 1772 voyage had begun by 30 May, when Captain Bickerton of the *Augusta* asked for an extra pilot for the trip and recommended Charles Rickersies.[6] A dozen men were taken from the *Marlborough* and iron ballast was brought on board – the trim of the ship would be a matter of concern throughout the trip. Sandwich came on board on the 8th amid the usual salutes and the ship progressed downriver.

The collier *Resolution* had been fitted out at Deptford for James Cook's second voyage of exploration. Joseph Banks, who had taken part in the first voyage, was a leader of high society in contrast to Cook's humble origins and campaigned to take charge of the next voyage. Appointment as a 'gentleman captain' might have been possible in the last century but could not be done in 1772, when the naval officer corps was well established, so Sandwich had to decline. Moreover, Banks insisted on the *Resolution* being fitted to his specifications with a superstructure which made her 'by far the most unsafe ship I ever saw or heard of', according to one of her officers. By the time she reached the Nore her defects were unmissable and soon, noted Midshipman Elliot, 'We were ordered into Sheerness, and the next morning at daylight, I believe two hundred shipwrights were cutting, and tearing the ship to pieces, in all parts of her, so much that it was dangerous to stir about … when this was done.' Banks and his party withdrew.[7]

The affair had clearly affected Sandwich. The *Augusta* came alongside her at the Nore on 10 June and his party went on board. He devoted more than two pages of his journal to praising her in her present condition; it was agreed by Cook and many others that she was 'in every respect adapted to the voyage she is going to undertake'. The *Augusta* weighed anchor and 'soon had the good fortune of a brisk gale at east, which brought us to Spithead the next morning by 12 o'clock.' After the usual round of inspections of ships and facilities the *Augusta* sailed on the 12th, in company with the *Terrible*, *Royal Oak*, *Centaur* and *Worcester*, which were going to join Admiral Spry for a summer cruise. There was drama as the *Centaur* ran aground on the Horse Bank outside the harbour but Sandwich noted calmly, 'as the weather was fine and the tide rising she floated off in about three hours without receiving the least damage.' They were visited by the Polish Prince Poniatowski, whose country was soon to be partitioned by Prussia, Russia and Austria. On the 16th they called for all available boats to help haul them out of the harbour. The voyage to Plymouth was slow. 'The wind being contrary we did not reach Plymouth till Sunday night' – the 21st.

On 2 July they were off Dartmouth on the return voyage and Sandwich noted: 'in the afternoon fell in with the *Resolution* sloop, went on board her, and made particular enquiry of the Captain and other officer, how she has behaved in her passage from the Nore, they all assured me, that they were perfectly satisfied with her, and that since the alterations made in her, there did not remain the least cause of complaint.' He went below decks to 'examine how the ship's company were lodged, and found they had as much room as they could possibly have occasion for, and more height than in many of our fifth rates, especially those of French construction.' Clearly obsessed with the issue, he filled more than two further pages of his journal singing the *Resolution*'s praises.

Progress was till slow and on Sunday 5th they were abreast of the Isle of Wight, where 'we were met by a strong gale of wind at east, and as we could not make any way in our intended course, it was judged to be a favourable opportunity to examine the six ships that are building by contract near Southampton.' The *Augusta* anchored in the sheltered waters off Calshot Castle at the entrance to Southampton Water, and used the ship's boat to visit the private Southampton shipyards and then passed through the New Forest with its potential supplies of timber, to look at the rural shipbuilding site at Buckler's Hard, where four ships were building for the navy. After these delays it was the 17th when they arrived at Sheerness, where Sandwich observed, 'the soil is so spongy being all made ground, that the water works through on the sides and it is a doubtful matter whether these inconveniences can ever be remedied'. Back at Woolwich, business was pressing and he returned to London to deal with it, coming back to Deptford a month later to complete his inspection.[8]

⁓ The naval review of 1773 ⁓

On 7 June 1773 Sandwich began that year's visit, which was to form part of a royal review of the fleet. 'This day I embarked on board the *Augusta* yacht and sailed from Deptford on my annual visitations of the dockyards, we proceeded directly to Portsmouth and anchored at St Helens in company with the *William and Mary* and *Fubbs* yachts, the *Hazard* sloop and *Greyhound* cutter.' They entered Portsmouth Harbour on the 11th and the *Augusta* hoisted the 'anchor and hope' flag to signify that members of the Admiralty Board were present. Sandwich would spend the next few days planning for a royal visit. He had already decided that His Majesty would sleep in the commissioner's house inside the dockyard, but hold levees in the governor's house outside. The King was beginning to show some enthusiasm for the visit, and wrote, 'My intention in coming to Portsmouth is to see the dock thoroughly, I shall therefore as one morning will not suffice, remain there till Saturday, you will make the arrangements accordingly.' Sandwich therefore planned a four-day visit, which the King considered 'very agreeable', and he looked forward to 'the many curious and entertaining objects I shall view.' Among his existing collection he found a 'clasped book of the navy' dating from 1698, which provided 'some amusement, and it may have given birth to some new ideas from seeing the improvements that have been made, and suggesting what further may be done.'[9] After Sandwich's usual inspections of the naval facilities, from the 18th 'every hour was taken up with making the necessary preparations for his Majesty's reception'. Some of the ships in Portsmouth Harbour were put into commission under Admiral Pye, while another squadron was sent up from Plymouth commanded by Admiral Lord Edgcumbe. They consisted of twenty ships of the line which were anchored in two lines at Spithead, plus two frigates and three sloops.

The *Augusta* was to be reserved for the King's use during the visit. In addition, the *William and Mary* was 'to receive and attend upon' Lord North, the prime minster, and members of the Cabinet. The *Fubbs* was to fulfil the same task for 'the Master General of the Ordnance, His Majesty's Privy Councillors, the Groom of the Bedchamber, the equerry and adjutant general, with any other persons his Majesty may allow to be admitted on board of her.' The dockyard yachts from Chatham, Portsmouth and Plymouth were available 'to receive such other of the nobility and persons of distinction as may be desirous of going to Spithead.'

The King arrived at the commissioner's house on the 22nd amid greetings from numerous political and naval dignitaries and the aldermen of Portsmouth. He was rowed out to the flagship *Barfleur* by barge, and on board he 'sat at a table of thirty covers, at which many of the nobility, and persons of distinction, as well as officers of the navy and army … were admitted to the honour of dining.' After that he boarded the *Augusta* to sail into the harbour with the usual salutes.

He was rowed round various ships the next morning and at six in the afternoon he boarded the *Augusta* again for his return to the harbour, but there was no wind. Sandwich was ready for that: the royal barge was on hand to take him. He spent most of the 24th in the dockyard itself, where Sandwich had arranged for him to view 'a principal piece of work of each branch'. Thus he saw the whole process of the making of a cable of 24in, the 'construction of a boat from the laying her keel to the finishing her for service', and the making of a large anchor, 'during which our royal master was very minute in his observations'. He announced several promotions and conferred knighthoods, including Captain Bickerton of the *Augusta*. He boarded the yacht again at five that afternoon to sail along the lines of ships at Spithead. As usual during the visit, the *Augusta* was accompanied by 'a very great number of yachts and other sailing vessels and boats, many of them full of nobility and gentry'.

The King was back in the yacht at seven in the morning of Friday 25th, when she sailed out of the harbour. When she reached the fleet at Spithead, Admiral Lord Edgcumbe's division set sail and followed the yacht as far as Sandown Bay on the Isle of Wight. The wind was now freshening and the tide had turned, so they proceeded back to the anchorage at St

Helens just inside the Solent, while Edgcumbe headed back to Plymouth. The *Augusta* took the King along the line of ships remaining at Spithead and he landed in the harbour, where he conferred more honours and promotions, including the lieutenant of the *Augusta* to captain, and two midshipmen from each of the ships and yachts to be made lieutenants. On the morning of Saturday 26th the King boarded his coach to return to Kew, having distributed largesse, including £150 to be shared between the crews of the *Barfleur* and *Augusta* and the royal barge. He was 'pleased to express the highest approbation of the good order and discipline of his fleet, the excellent condition of the dockyards, arsenals and garrison; and the regularity with which every thing was conducted'.[10]

Sandwich, however, was not finished with his travels. He sailed to Plymouth in the *Augusta* on 30 July and arrived after a fast voyage of twenty-four hours in a brisk easterly wind. His return began on 6 July but was much slower and it was not until the 12th that he arrived at Chatham, noting the common difficulty in getting ships up the Medway against the prevailing winds. He sailed to Sheerness. He was at Woolwich on 16 August and then Deptford on the 17th, where he noted the state of the remaining yachts with his usual eye for economy.

> The repairs of the *Royal Charlotte* are now completed, and when the gilding is finished she will be ready for sea; the *Katherine* and *Mary* yachts are under a thorough repair, or rather rebuilding, on two of the slips adjoining to the basin, and when they are finished, as there will be three yachts besides the *Royal Charlotte* including the *Princess Augusta* in complete repair, and little probability that more than that number can be wanted for service, it seems to me that the others may be suffered to decay, by which a considerable and unnecessary expense may be saved to the public.[11]

The *Fubbs* did not have so much as a small repair after that and was scrapped in 1781. The *William and Mary*, however, was recommissioned in 1779 and survived until 1801.

1774

On 4 June 1774 the newly renamed *Princess Augusta* received Lord Sandwich's stores, and he came down to Deptford to embark two days later, to be saluted with eleven guns. Sandwich noted that 'the *Augusta* yacht has also been entirely rebuilt and altered from a ketch into a ship of three masts.' She had 'changed her name to the *Princess Augusta* and was become a post ship, a distinction very justly due to her, in consequence of her having had the honour of carrying her royal owner to sea at the time of the naval review last year at Portsmouth.'

They arrived at the Nore that evening to be joined by the *Hazard* sloop and *Greyhound* cutter. They were cheered by the crew of the guardship at Sheerness and headed up the Medway to Chatham, where there were 'no great works carrying on'. They took a few hours to sail to Sheerness but only planned a short stay as the yacht used only a single small anchor, the kedge, to hold her fast. They soon raised it and went on as there was 'very little alteration made since the last year'.

They were rejoined by the *Hazard* and *Greyhound*, but the passage onwards was slow, for unlike the King they could not choose the time of their passage to take advantage of favourable winds. They had to anchor at Margate Flatts twice, and were towed part of the way by their boats. They rounded South Foreland near Dover, but in the Channel they remained in sight of Beachy Head for two days. The tall spire of Chichester Cathedral came in sight in the morning of the 14th and finally they made fast to a buoy in Portsmouth Harbour next day. To Sandwich, new wharves, part of a 'great plan of improvements', offered more of interest.

They sailed again on the morning of the 20th, passing ships at Spithead and accompanied by the *Greyhound*, *Hazard* and Lord Ferrer's yacht. Progress was still slow in unfavourable winds; they had to anchor off Newton Harbour and Yarmouth before passing the Needles into the English Channel. By the 24th they were 'tacking occasionally' off Berry Head in strong west-southwesterly winds which obliged them to lower the topgallant yards. They got to Plymouth next day to see more wharves and other works which were 'going on with all possible expedition'. Meanwhile, the yacht was heeled and her bottom scrubbed in an attempt to increase her speed. As always during the voyage, they took on stores from the shore: 'received six bags of bread and one quarter of fresh beef and some pork'. They sailed again on 3 July and stopped at Portsmouth again for two days. The winds were light and they had to use studding sails to take them past the familiar landmark of Beachy Head, but they were off Margate by the 8th and after a stop at Woolwich were back at Deptford by the afternoon of 12 July.[12]

1775

Joseph Banks came on the 1775 trip, apparently reconciled to Sandwich after the dispute over the *Resolution* – though he never missed a chance to sneer at the navy's seamanship. On 2 June the party, consisting of Sandwich, the Earl of Seaforth, Omai, a Tahitian brought over by Cook, and two others, 'met at the Admiralty and after breakfast proceeded to the Tower, where at 8 we embarked on board the Admiralty barge and proceeding through the forest of masts always to be found in that part of the Thames, … passed the windmills on the Isle of Dogs'. At Deptford they boarded the *Princess Augusta* under the command of Sir Richard Bickerton, but off Greenwich:

The royal visit to the fleet in June 1773, with the Royal Standard flying from the *Augusta*, by Dominic Serres. (Royal Collection Trust / © Her Majesty Queen Elizabeth II 2022, 404559)

> our yacht fell desperately in love with a Dutch vessel in spite of all our efforts to prevent her proceeded to such familiarities that Sir Richard ordered her to an anchor and as a punishment for her libidinous inclination and effectually to prevent the generation of so foul so monster as would have sprung from such illicit amours confined her during the rest of the ebb.

Bickerton made no mention of this in his log.[13] They visited Chatham and Sheerness again and arrived at Portsmouth on the 13th after a slow passage. On the 25th they passed through the Needles on the way to Plymouth, where they arrived on the 27th.

The return passage encountered some bad weather, and on 10 July at the mouth of the Thames Estuary the *Princess Augusta* lost a flying jibboom. Back in the Thames in the 12th:

> … yacht sailed for Deptford where she presently arrived but no sooner did she find herself in her own dominions than tired of her voyage I suppose and willing to take a little rest she incontinently ran herself ashore upon a mud bank not very far from the place where she had before ran foul of the Dutchman. Sir Richard swore, the pilot stormed but no reason would she listen to nor any argument but that of a hawser which being hove taut taught her to behave herself better, and indeed she remained very quiet at anchor all that even.

Meanwhile, Bickerton raised expectations of a meal on 'the most exquisite round of beef' and 'our appetites were worked up to a pitch of expectation', but no beef appeared for dinner – instead at supper they had 'a quantity which would have done credit to a troop of his majesty's beefeaters.' Next morning was 'the melancholy day of parting', for Banks was determined to avoid Sandwich's muster of the yard – at another yard he had noted, 'as Lord Sandwich had destined this day for mustering the yard, a matter of infinite consequence to him though of no amusement to us'. Banks had mixed feelings on parting, which he disguised with irony – 'I confess I have for the uninterrupted pleasure I have enjoyed during the whole of this voyage will upon a comparison destroy the effect of any which even the capital can furnish. I must therefore content myself with saying and thinking that I have been happy.'[14]

This was the last of Sandwich's visitations. Already the American colonists of Boston had staged their famous tea party and coercive acts were passed against Massachusetts. The Continental Congress had met in Philadelphia and more recently British forces were defeated at Lexington and Concord. A new war was about to begin, which would drag in France, Spain and the Netherlands and ruin Sandwich's project to reform the dockyards, as well as his reputation for many years.

3
An Unhappy Marriage

Joseph Goodall perhaps hoped to see a little of the world when he formed part of a group of ten members of the royal household who left St James's Palace, London, on Sunday, 8 December 1794. Goodall was one of the 'silver scullery' team in the palace, charged with cleaning the most expensive of the royal plates after meals.[1] They were to travel to Deptford to join the *Princess Augusta,* which was ordered to cross the North Sea to pick up Princess Caroline of Brunswick and bring her back to marry George, the Prince of Wales, an arrangement which would cause the prince's extensive debts to be paid.

There was a setback immediately as the barge they were supposed to join at Whitehall stairs missed the tide. They went back to the palace and then on to Deptford the next day. They boarded the yacht and waited nine days for a fair wind, but it still took them three days to reach Sheerness. Goodall was impressed with the 'several noble ships' he saw along the way, and with the great congregation of gulls at Sheerness. He messed in the gunroom with his colleague Richard Miles, Mr Boyd, the ship's acting steward, Rowe, the captain's servant, Edderly, the master at arms, Morrison, the 'assistant in the gun room', a boy to wait at the lieutenant's table, and John Murry, 'a black, he plays French horn'. He was thirty years old and a native of St John's, Antigua.[2]

They remained at Sheerness awaiting news of the princess's movements. War with revolutionary France had begun nearly two years earlier. Allied armies were being driven back on several fronts, including the Netherlands, but the Royal Navy had won a great victory on the Glorious First of June, 1794. From Sheerness the crew and passengers on the *Princess Augusta* could see a stream of ships sailing to join the war. The weather worsened. On 29 December the wind was 'very blusterous and the water very rough'. The ship drove her anchors and was in danger of being blown out to sea until her cable was cut. It happened again on the 23rd and Goodall, who always took an interest in the running and organisation of the ship, noted, 'we were forced to have all hands aloft and with a deal of labour once more safe in Sheerness, the wind very blusterous and quite contrary.' It happened again next day but they returned to Sheerness for Christmas, which, however, brought no relief. On the 30th they were visited by Lady Jersey, the prince's mistress, and Lord Clermont, his racing crony, who stayed for a few days. On the first day of 1795 they had to put into Sheerness for provisions, for yachts were never intended for long voyages.

They lay at Queenborough on the Swale for days with fair weather but no orders. On the 8th 'received on board by the *Duke William* lighter from Chatham ten -?- of water, two hogsheads and ten half hogsheads beer; returned by the same conveyance four empty puncheons, one empty hogshead, three half hogsheads all full hooped, 72 staves and 132 hoops.'[3] On the 18th the weather turned very nasty, with frost and snow. It was even worse on the 20th being 'very frosty and bitter cold and such surprising large flecks of ice scarce even known.' Supplies were running short again; they were hit by a drifting Dutch ship which damaged the bowsprit and took away part of the figurehead. On the 24th, according to her master, 'got our stream cable on shore and hove by it till we come aground'. Early next morning, 'being high water hove the ship nearer to the shore that she laid more on the flat; at 9 AM got our stream anchor on shore and at high water hove the ship still nearer that when she grounded the ship laid much on account of a flat.' They placed planks over the ice to allow the crew to get ashore for more provisions.

A thaw began on the 27th with 'large heaps of snow swimming about us', but it was a false hope and the frost was back by the 29th. Ten pressed men had been drafted from the *Sandwich*, the flagship and depot ship at the Nore, and two of them took the opportunity to desert. Lieutenant Seymour and the pilot went after them, but William Carter got away. Francis Kimber, a 23-year-old seaman from New York, was apprehended and put in irons, but evidently he was forgiven – he was not sent for court martial and there is no record of him being flogged.[4]

On 2 February there was a slight let-up and fishing smacks began to set out, while workmen arrived from Chatham or Sheerness to repair the bowsprit and figurehead. Goodall was allowed ashore and found Queenborough to be 'a neat, clean place', sending two representatives to Parliament, though it only had one street. On the 18th it was cold again, with another hazard: 'our ship was set on fire in the galley by a sailor who was drunk, he took his pipe into his cabin and set fire to his clothes and burnt the bottom of his bed, it was safely put out in good time.' By the 20th they were all 'very impatient for orders, most of us out of all patience.' Even a royal yacht was not immune from the attentions of the press gang, who came on board on the 23rd, 'to see if we had any men to spare but they took none.'

On the 26th there were indications of a change of plan. The princess was to come over on a larger ship, the 50-gun *Jupiter*, and a pilot and two of the kitchen staff were transferred to her. The *Jupiter* finally sailed in a southwesterly wind on 2 March and the *Augusta* had orders to await her return. By the 6th they were at Sheerness again and Goodall observed, 'the sea this time is beautiful to behold'. On the 14th they were 'expecting every hour to see the *Jupiter* arrive.' Next day there was drama as the *Porcupine* frigate ran aground and was briefly in danger from the enemy. On the 16th Goodall recorded, 'great preparations is making at Greenwich for the reception of the Princess of Wales elect. *The Princess Augusta* yacht rolls about nicely'. But Goodall's journal continued wearily to record 'Still at the Little Nore'. He amused himself by observing the passage of ships on the way to the wars: on the 18th he saw 'fine frigates passing', on the 24th there was the *Diamond*, 'a fine fighting ship', captained by the hero Sir Sidney Smith. But by the 28th there was still no news.

For the *Jupiter* and her consorts had their own troubles, first with fog and then with ice floes coming down the River Elbe sweeping away the marker buoys, and severely damaging the *Jupiter* on 11 March. The princess stayed at home in Brunswick, for the roads were impassable, and it was the 28th before she boarded the *Jupiter*, which sailed the next day.

Finally after twenty days at the Little Nore, the *Princess Augusta* received word that the princess was only twelve miles away. She moved up to the Great Nore and among the weary crew, 'all hearts now appear to be rejoiced.' The *Jupiter* arrived and Captain Browell was called on board her at eight in the morning of the 4th and was ordered to accompany her and the *Mary* yacht upriver to Gravesend. Meanwhile, Lord Malmesbury (who had negotiated the marriage, but now had doubts about the princess's suitability), Commodore Payne and the captains of the *Jupiter* and *Mary* came on board the *Augusta* to view the accommodation, and Captain Browell was invited to dine with the princess on board the *Jupiter*. When they sailed the *Mary* took the lead until she was signalled to keep her proper station, 'at which her captain was not best pleased'. They spent the night anchored at Gravesend. At last, at 8.30 on the morning of Easter Sunday, 5 April, came the long-awaited moment when the princess boarded the yacht, as described by Goodall:

> Her German attendants came first with three other servants one woman and two men in a fine gilt boat and then came the Princess in the Admiralty boat all most beautifully gilt and all the rowers, 12 in number with a coxswain, all dressed in new scarlet with the heavy badge of silver on their arm the coxswain's badge was gilt with gold of a gold lace hat, and a fine canopy over the princess and her company. Lieutenant Mainwaring steered the Princess to our yacht. The boat she came in was ornamented with a beautiful flag at the head and the moment she boarded us we hoisted the royal standard of England at our mainmast. Wind was very fair we set sail directly and a very fine day's sail we had and thousands of spectators each side to view us.
>
> The company that accommodated the Princess was E[arl] of Malmesbury, Commodore Payne, Lt Mainwaring
>
> The Admiralty boat with several cutters did accompany the Princess up the river, the people and boats was joined all the way up the river and guns fired from every ship and cutter that had got one. We landed the Princess at Greenwich at half past one o'clock and thousands of people was present …

The princess was on board the yacht for just five hours, after it had waited more than three months in very unpleasant conditions for her to arrive. The Prince of Wales was not pleased with his bride, especially her irresponsibility and her failure to wash properly. It was too late to cancel the wedding, and he was visibly drunk during the ceremony on 8 April, three days after her arrival. They stayed together long enough for her to conceive a daughter, Princess Charlotte, and after that Caroline lived her own life, bringing near disaster to the monarchy.

The quotes from Joseph Goodall are from a diary entitled *Small Observations on our Voyage to Fetch the Princes [sic] of Wales*. I am grateful to the owner of the diary, in private hands, for allowing me access to it, and to Kevin Fewster and Stephen Riley for drawing it to my attention.)

Princess Caroline landing at Greenwich from the Admiralty barge. (Royal Collection Trust / © Her Majesty Queen Elizabeth II 2022, 605313)

~4~
The Yachts at Weymouth

Late in 1788 King George was taken ill, with symptoms that were taken for madness at the time and resulted in the employment of a straitjacket and restraining chair. He was beginning to recover by next spring but was still weak, and the question of his convalescence arose. His brother, the Duke of Gloucester, had first discovered the resort of Weymouth in 1780 and had a house built on the still vacant seafront there. Dr John Crane of the town described the resort's virtues. The greatest single advantage was that its bay was protected from westerly winds by Portland, allowing smooth waters for bathing and sailing. Unlike Margate, there was no major river flowing by, bringing sewage and diluting the seawater with fresh. And unlike Brighton, its fishermen had the use of a harbour at the mouth of the River Wey, so there was no need to haul their boats up or stretch out their nets on the beach. Moreover, the King's estranged son, George, Prince of Wales, was already established at Brighton

The royal party travelled down by coach over five days at the end of June. Once there, it is not known if King George drank seawater as Crane advocated, but he certainly bathed regularly. He took to the water twenty-one times during the six weeks of the first stay in 1789, even though that was interrupted by a two-week tour of the West Country. It was always done as Crane recommended: 'The most proper time of bathing, is early in the morning; before which, no exercise ought to be taken … To bathe late in the day (more especially in hot weather) will occasion great depression of the spirits, particularly in debilitated, or paralytic persons.'[1]

Crane also recommended sea voyages – as a naval surgeon escorting ships to Lisbon in 1760, he had seen miraculous recoveries by passengers. These too would form part of the King's regime. Apart from the Prince of Wales, the King's six surviving sons were officers in the army and navy and were rarely able to attend at Weymouth. The six princesses, however, had time on their hands and none was married until the eldest, Princess Charlotte, was wed to the unattractive Prince Frederick of Wurttemberg in 1797. Mostly they were not keen on Weymouth and Princess Mary wrote in 1798: 'after all the amusement we have had at Windsor this place appears very dull, indeed stupid and samey.'[2] However, they did not reveal their dislike to their father, who was kind and loving in his way, but was not easy to communicate with.

Weymouth in 1803, by John Thomas Serres. (Royal Collection Trust / © Her Majesty Queen Elizabeth II 2022, 701407)

In 1789 the Admiralty appointed two ships to attend the royal family at Weymouth, 'for the security at once and the pleasure of the King'. The *Magnificent* was a 74-gun ship of the line, though it was quite rare for such large ships to put to sea in peacetime. Her captain, Richard Onslow, was the senior officer of the guardships at Portsmouth, which was probably why she was chosen for the task. The *Southampton* of 1757 was the first of the 'true' frigates with a single gundeck and unarmed lower deck, a classic type which Nelson, for one, craved more of. As an older ship, she could be spared from active duty to attend the King.

The royal party went on board the *Southampton* on 5 September. According to Miss Georgina Townshend, the princesses had never seen a ship before and 'I never saw people enjoy themselves more than they did', even though they had to be hoisted up the side, which 'entertained the King very much.' They boarded the *Magnificent* the next day, Sunday, attended by various nobles, and they were impressed. The Queen reported:

> We have therefore begun sailing as a substitute for riding. The King has got the *Magnificent*, a man of war of 74 guns, and the *Southampton* frigate of 32 in the Bay; in the latter we have sailed yesterday for the first time, and we row frequently in ten-oar cutters. Our rowers are those of the Duke of Clarence; they are very smart fine dressed men.[3]

Big ships like the *Magnificent* were not sent again, but the *Southampton* became a regular visitor.

Exposed on the south coast, facing French territory, Weymouth clearly had its dangers after the declaration of war in 1793, as the courtier Fulke Greville observed. 'The situation of Weymouth, the tides in the bay and neighbourhood, as well as a period of war, of course excited prudent attentions when their Majesties went afloat.' There was a scare in September 1794 when the frigate *Trusty*, having encountered four French frigates in the Channel, entered the Bay with signals of an enemy sighting still flying. All was well, next day the King witnessed the manoeuvres of Saumarez's small squadron, and 'expressed himself much satisfied'.[4] Later in that eventful month a brig was driven from her moorings in the Bay by 'the violence of the wind' and fired distress guns. She was saved with the help of the *Southampton*'s boat.[5]

Yachts were not yet used for the King's entertainment and health at Weymouth, but the patterns were established. The ships' logs record the arrivals and departures of the royal family, but are vague about the details of where they sailed – the waters were familiar to the officers, had no hidden hazards and they needed very little navigation. During the first trip in 1789, the Queen recorded, 'We sailed back and forwards in the Bay'. The courtier Fulke Greville gave a more general description of their moderate cruises which were generally between St Aldan's Head and the Bill of Portland:

> The Bill of Portland varied as circumstances of wind and tide permitted. St Aldan's Head is seven leagues from Weymouth, and on those trips it was approached or fetched in varied directions, either from more open sea or by sailing nearer to the chalk cliffs by Lulworth, in passing which the look into Lulworth Cove is very interesting, opening Lulworth Castle gradually to full view. …
>
> In favourable weather his Majesty's cruise was occasionally extended into the West Bay, beyond the Bill of Portland. Sometimes the sailing squadron stretched further out into the Channel, and particularly when homeward or outward bound fleets were passing.

On 11 September 1789, according to the Queen, 'We sailed considerably beyond Portland, saw Lord Westmorland to accomplish to get up to the foretop. Dined on Board about 2 … a remarkable pleasant day.' And there was another distraction:

> We saw there a fish catch by one of ten sailors called a sculping. It is not large in the body but has an immense mouth and large rows of teeth, in the sides two pouches where it lodges its young ones, and two stumps in the shape of a hand between the throat and stomach to take hold of them, it is very unusual in this part of the world, frequently taken in North America.

The following day there was a different kind of amusement. 'We saw the son of the surgeon of the *Southampton*, Mr Fig Morrice (aged 5 years) to go up as high on the foretop, which he did incomparably well.'

Sometimes they sailed out a short distance to meet a passing fleet or convoy – Admiral MacBride's in August 1794, Sir John Borlase Warren's in September and Admiral Howe's later in the month. That year they made their first visit to the 'island' of Portland, which was long remembered as a special event:

> The King was received by the inhabitants of the village of Chiswell with music playing … then went all round the island, saw the old and new lighthouse, the latter of which is 75 feet high just built, then returned to Chiswell and dined at the Portland Arms … about four went into the *Southampton*'s barge to row home …[6]

Sailing was often postponed for one reason or another. The Queen wrote in August 1797, 'our sea excursion have not begun as the other frigates were not come into harbour' and five days later after they did, 'This week was intended for sailing chiefly, but I believe in that amusement we shall be disappointed for at present it blows a hurricane, and should the wind fall there must remain a swell in the sea which may probably last several days.'[7] Her Majesty put a brave face on it and suppressed her memories of her horrific voyage in 1761, but she only went out of loyalty to her husband. In September 1798 Princess Mary wrote:

> Mama, I fear, is beginning to *feel* unwell as she always does whenever she is at Weymouth, but she assures me she feels better today than she did yesterday evening. One thing makes her very happy which is its being *determined* that the *sea* parties are not to take place. This year Lord Spencer has sent down a barge and whenever the weather will permit we are to row, but *entre nous* mama is so much afraid of any *motion* that I do not think papa will get her to go.[8]

The Queen herself tended to prefer calm weather. In October that year she wrote, 'At half past ten we went on board the *St Fiorenzo* … We really did breathe the finest sea air imaginable, but there being no wind whatever we enjoyed the sight of all the little pleasure boats rowing round the ship and hearing all the loyal and sea songs from all quarters.'[9]

The King was well aware of the naval mutinies at Spithead and the Nore in 1797, if only because he had to confirm death sentences on some of the participants in the latter affair. The hero of the Nore mutiny was Harry Burrard Neal, captain of the frigate *San Fiorenzo*, which had been at Corsica in 1794. He retained the loyalty of his crew and escaped under fire from the anchored ships, a turning point in the mutiny. In 1798 the King asked specifically for the *St Fiorenzo* to attend at Weymouth. She already had a reputation as a good sailer.[10] From Worthing, more than a hundred miles away, Princess Amelia commented, 'I am very glad the *St Fiorenzo* is at Weymouth as I know how much you all like sailing in her. I shall ever look back with pleasure to those happy hours I have passed on board. I know how happy Sir Harry is to be on that station.'[11]

∼ Camperdown ∽

That year the King had an all too brief excursion in a yacht. Admiral Adam Duncan defeated the Dutch fleet at Camperdown on 11 October, a decisive and much-needed victory in a grim year. The King decided to go down to meet the triumphant fleet at the Nore and the yachts *Royal Charlotte*, *Princess Augusta* and *Mary* were ordered to be got ready. The King arrived at Greenwich on 30 October and the pensioners, 'maimed and worn-out British seamen', were lined up for the royal carriage to pass between them. He had an 'elegant dejuné' with Lord and Lady Hood before embarking in the royal barge steered by Captain Trollope, as the tide now served. He boarded the *Royal Charlotte* to the sound of a royal salute while the Lords of the Admiralty, led by Earl Spencer, boarded the *Princess Augusta*, and the members of the Navy Board, the *Mary*. There was a slow passage to Blackwall but the wind served better to Woolwich. According to a newspaper report, 'The greatest number of boats ever seen on that part of the river, attended on the occasion, from a Margate hoy down to a Westminster peter-boat, all full freighted with passengers and provisions'. They got as far as Lower Hope downriver from Gravesend, but a northeast wind blew for several days 'with uncommon fury'. The King, it was reported, had 'shewn himself the oldest and ablest seaman of any of his noble attendants, some of whom have been quite seasick, while the King has weathered the storm with high health and spirits.' He was keen to continue the voyage, but Captain Trollope and the other officers persuaded him against it. Apart from the problem with the weather, Parliament was due to meet in a few days and there was no guarantee of getting back in time. The *Royal Charlotte* arrived back at Greenwich ahead of the *Princess Augusta* and *Mary* which were 'somewhat heavier sailers', to be greeted with more salutes and the King took a post-chaise to London. He hoped to return another day, but never did.

∼ The yachts at Weymouth ∽

From 1801 the family's nautical excursions off Weymouth were mostly done in yachts rather than frigates, perhaps at the instigation of the new First Lord of the Admiralty, the forceful Earl of St Vincent. In May he wrote, 'Lord St Vincent has had the honour of receiving your Majesty's commands for preparing the *Royal Charlotte and Princess Augusta* yachts for the purpose of attending your Majesty during your intended residence at Weymouth, and will give the necessary directions immediately.'[12] In July he was 'very happy to learn the *Royal Charlotte* and *Princess Augusta* yachts answer your Majesty's expectations – a great deal depends on the good management of their commanders.' And in view of the invasion scare of the time he found a new use for the remaining yacht the *Mary*, or perhaps revived an old one. She was to be commissioned and form part of a force guarding the Thames Estuary under Lord Nelson.[13]

By 21 June the two yachts were running down the River Thames in an easterly wind, so they depended on the tide to carry them on. Both ships anchored at Gallions Reach in the afternoon as the tide turned, to set sail again at ten next morning. They reached Gravesend around four and anchored again. A brig came athwart the hawse of the *Princess Augusta* and carried away her flying jibboom in the bows, which tore down the fore topgallant mast. Nevertheless they set sail again next day. There was an exchange of men with other ships on the 24th while anchored near the Buoy of the Nore. Next day they passed round the North Foreland on the northeast extremity of Kent. The wind was veering to west-northwest and there was no need to stop in the Downs. It turned more easterly again so they turned down-Channel, passing Dungeness, Beachy Head, Shoreham and the Owers Bank to enter the Solent and anchor at Spithead. As soon as the tide was favourable they tacked down to Lymington where they anchored to await the royal party. *Princess Augusta* took the opportunity to repair the damaged rigging. Stores were taken on from Portsmouth Dockyard and crews were employed cleaning and painting. A cutter arrived with Burrard Neale, who was to take charge of the operation. At dawn on 1 July they weighed again and sailed through the Needles Channel to anchor in Christchurch Bay, where they endured 'great falls of rain' overnight.

Normally the royal party travelled by road to Weymouth over several days. In 1801, however, the royal party stayed in Cuffnells, the house of George Rose, a former Chief Secretary to the Treasury, near Lyndhurst, which the Queen preferred to Weymouth.[14] At Christchurch they embarked in the royal barge to be rowed to the *Royal Charlotte* and the squadron, including the *Fortune*, *Hydra*, *Sheerness* and two cutters, saluted with twenty-one guns. They sailed but the wind changed and the Queen lamented a 'sailing of very near 14 hours' to reach Weymouth early next morning, where they anchored offshore and the passengers disembarked to take up residence in Gloucester House. The town was illuminated, and rockets were fired. Then the yachts moved to an anchorage further inshore – one of their advantages over the frigates was that they could get closer to the house and shorten the distance to be rowed, and by the harbour mouth they could get more shelter from southerly winds.

Over the next nine days there were 'strong' or 'fresh' gales which prevented any aquatic activities, causing the *Princess Augusta* to lower her topgallant masts for safety for a time. The crew of the *Royal Charlotte* took the opportunity to conduct a survey of the entrance to Weymouth Harbour, which they would find useful later. At last at 10am on the 15th the royal family embarked on the *Charlotte* to the usual 21-gun salutes. With the squadron she stood: 'spent the day standing off and on in Portland Road ... at 2 pm brought the ship ahead inshore, at 4 wore ship and stood off under the topsails, at 5 tacked in shore.' The family disembarked at six according to the log of the *Augusta,* and 6.30 according to the *Royal Charlotte* – such discrepancies were common in logs of the

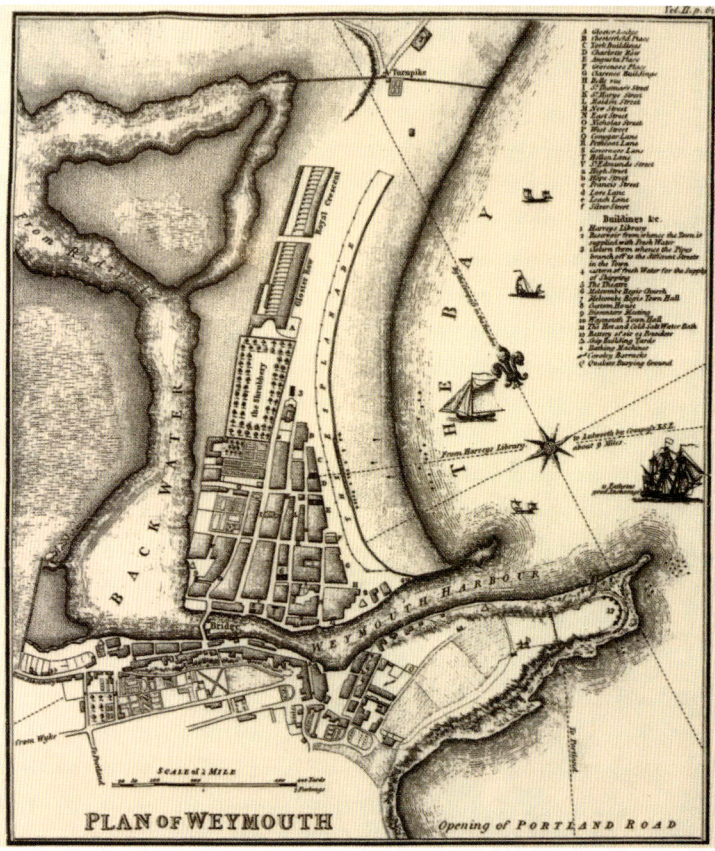

A plan of Weymouth and the Nothe peninsula in 1800. The royal yacht probably anchored close to the compass rose marked in the bay. (Weymouth Museum)

period. Sailing was cancelled the next day due to 'strong winds and rain' and it was the 20th before the royal family embarked again, to sail out towards Portland. There were two more sailing days before the winds rose again, but it resumed on the 27th.

During August the yacht crews were involved in shore-based activities. Men from the *Princess Augusta* were landed to help erect marquees and adorn them with flags for a fete given by the King at Radipole to the north of Weymouth – as sailors they were used to handling ropes and canvas. Sailing activities continued when the weather allowed, and at the end of the month both the yachts entered the narrow waters of Weymouth Harbour, presumably reassured by *Charlotte*'s survey. They moored alongside one another and began to prepare for an evening 'entertainment', receiving 'fruit and viands etc from the Lodge' and dressing the ships overall with flags. At two o'clock the royal family and 'a numerous party of the nobility etc came on board'. About eighty of them were served dinner on the *Augusta* before disembarking at seven, leaving the crews to clean the ships. The routine continued for another month until the beginning of October, when the royal family left Weymouth by coach and the yachts began to return east. They made a fast passage and were soon in the Downs, where they returned crewmen lent from other ships. They were back at Deptford by the 5th, where they entered the wet dock and

THE YACHTS AT WEYMOUTH

began 'dismantling' or unrigging. Then it was back to the usual port routines, where the weather was the most interesting item recorded in the logs.[15]

The *Royal Charlotte* and *Princess Augusta* set off again in June 1802 but only arrived at Weymouth after a seventeen-day passage against westerly winds. Sir Harry tired of waiting for them at Lymington and pressed on to Weymouth. The royal family came on board as soon as they arrived, and began the usual round of trips towards Portland, Lulworth and round the bay. The only unusual event was in the morning of 13 August, when they witnessed an assorted collection of thirty boats having a sailing match. In 1803, with the resumption of war and a serious threat of French invasion, they spent their summer at Lymington instead, which must have pleased the Queen.

In 1804 they had the use of the new *Royal Sovereign* for the first time, and on 18 August she received orders to proceed there. Her topsails came on board and the courses were bent, she was hauled out of the dock and alongside the *Woolwich* of 44 guns before setting sail on Tuesday 21st. She passed Lymington and the Needles and dropped anchor in Weymouth Bay on the 24th, in time for the royal family to come on board two days later. They were accompanied by the King's sons, the dukes of Cambridge, Kent and Cumberland, six lords, three generals and eight others. For entertainment they had the twenty-four members of the 'Queen's band of musicians'. On

A visitor's guide to the panorama in Leicester Square, London – the viewer stood in the centre of the display. It shows the *Royal Sovereign* in the top right-hand corner, with the royal barge approaching it. (Weymouth Museum)

other occasions they listened to the 'German band of musicians', ten of whose eighteen members came from the Kellner and Greisbach families.[16] Next day the yacht took the royals round the bay towards Portland. There were similar trips on thirty days during the next two months, mainly interrupted by bad weather. On 29 October the *Royal Sovereign* set sail in the company of the *Crescent* and *Aeolus* frigates and the yachts *Royal Charlotte* and *Augustus*. After a stop at Lymington they were back at Deptford by 11 November, where the *Royal Sovereign* returned stores in anticipation of a quiet, and as it turned out frosty, winter. It was a similar story in 1805. The *Royal Sovereign* sailed from Deptford on 9 July and anchored under Portland Castle on the 12th. The royal family was on board for thirty-one days between then and their departure on 4 November. From the ship the King wrote to the Bishop of Worcester, who replied, 'I am happy to know your Majesty was employed in one of your favourite excursions, and that there seems reason to apprehend from the state of your eyes and the report of Mr Phipps that the perfect cure may be reported.'[17] But it was the King's last visit: his declining health prevented any more.

5

New Yachts for a New Century

The *Royal Sovereign*

The use of the yachts at Weymouth had perhaps revived royal interest in them. At the resort in 1801, the old *Royal Charlotte* and *Princess Augusta* were found to have 'defects to make good', especially the latter which was found to be leaky.[1] A new yacht was designed by Sir John Henslow and ordered to be built at Deptford, where the draft was sent at the end of 1801. She was 6ft longer, nearly a foot broader and 46 tons larger than the *Royal Charlotte*. Main compartments included the fore, middle and after cabins, the lords' room and the gentleman pages' room, captain's cabin, officers' cabin, gun room and crews' quarters. She too was to be ship rigged, with six round gunports towards the bows on each side and four square windows or 'lights' in the royal compartments, though in many ways she followed the style of the older vessel.

She was begun at Deptford in November 1801 after the Treaty of Amiens ended the war with France, for a time at least. There was concern about finding suitable timber for such a special vessel. The material left over from Sir Samuel Bentham's experimental vessels *Dart* and *Arrow* was examined at Redbridge in Hampshire and some was useful; there was more at Chatham, and Tippett the master shipwright at Deptford particularly requested 'that not one piece which will do for her will be used for any other purpose'.[2] The Thames shipyards were searched, John Dudman's offered nothing, as did John Wells, while Brent offered to find some on his next visit to the country. By mid January 1802 the Navy Board was doubtful about progress and ordered the yard to 'discontinue employing any men on the works of the yacht until the materials are provided.' The yard officers reassured them:

> that from the timber now landing from Redbridge, with a number of filling timbers and the keelson pieces that we have to trim, will enable us to keep the present people employed on the yacht, until the timber we expect from Mr Ellis from Whittlewood and Salcey forests arrives, which we are in hopes will nearly complete the frame …[3]

The *Royal George* at anchor in Cowes Road off the Isle of Wight *c*1819, by J C Schetky. (Royal Collection Trust / © Her Majesty Queen Elizabeth II 2022, 920266)

Despite this the shipwrights were soon complaining about low earnings due to shortage of material and the low task (piece) work rates paid for such a ship. The officers were sympathetic and asked 'that the people employed in building the yacht may be allowed two for one to the present time, and that they may be allowed the prices of articles for a 28-gun ship on carrying on the remainder of the works.'[4] In February 1802 the officers reported that 'the frame of the new royal yacht building here is nearly in such a state of forwardness that the joiners may prepare for framing the principal thwartships bulkheads, also the frames of the bulkheads for the officers' cabins'. But the work was 'very extensive' and a great part was of mahogany which would require 'the greatest care and attention'. At least twenty-four joiners were needed, but in the reduced post-war circumstances the yard only had eleven with enough skill, so more were to be brought in from Woolwich and a few house carpenters were to be employed, working from six to six on task each day.[5] They duly came, but in August they complained about the low rates of pay.[6] One of the painters of the yard was 'equal to the performance', but he left and another had to be sought in London.[7]

The new yacht was named the *Royal Sovereign*. Her launch was described with a professional eye in the *Naval Chronicle*. 'At a quarter before three o'clock on Saturday, May 12th 1804, the new yacht built on purpose for His Majesty was launched from the King's Dockyard at Deptford. She is a very neat but small ship. In her present trim she draws about 9 feet forward and 10 feet abaft. She is completely copper bottomed'. Her hull had cost £6537 for materials and £4730 for workmanship. Her masts and yards were remarkably cheap with £67 for materials and £147 for workmanship, which suggests that older ones from smaller warships might have been reused. In contrast, her rigging and stores cost £6301 for materials and £47 for work.[8]

She served the King at Weymouth that year:

> Upon one fine morning, in the summer of 1804, when his Majesty was on board, this yacht left Weymouth Roads and proceeded on a cruise, accompanied by the *Royal Charlotte* and *Princess Augusta* yachts and a frigate. The new yacht excelled her companions so much in point of sailing as to drop anchor in the Roads upon her return at six in the evening, while the *Royal Charlotte* did not arrive until ten o'clock that night, and the *Princess Augusta* until six o'clock the next morning – an unquestionable proof of the very great superiority of the *Royal Sovereign*, a superiority which gives her the eminent distinction, beyond controversy, the best sailer in the Royal Navy.[9]

Lieutenant John Boteler joined her at Deptford in 1822. At sea, he found she had 'the character of being, and was, a most excellent sea-boat, but she did kick and roll about famously.' Later, 'It was blowing very strong and we had to beat from the Nore into Sheerness Harbour; the yacht worked like a top, tack and tack without a fault.'[10]

～ The *Royal George* ～

On 20 April 1814 the Admiralty ordered a new yacht and her draft was prepared by Sir Henry Peake, the current Surveyor of the Navy. On 4 June the Navy Board approved 'A draught of a yacht prepared in pursuance of an order from the Right Honourable the Lords Commissioners of the Admiralty' and in turn the Admiralty approved it on 4 June. It was sent to Deptford 14 July 'for building the *Royal George*'. She was more than 7ft longer than the *Royal Sovereign* and a foot broader, measuring 330 tons, an increase of 25. In overall dimensions she was not unlike a small warship, coming between the 454 tons *Cyrus* class of ship sloops of 1812 and the 235 tons of the *Cherokee* class of brig sloops of 1807, though she was narrower than both in proportion to her length, and deeper than the *Cyrus*. Her sides were pierced for nine gunports on each side, but there was a substantial gap between the aftermost port and the quarter gallery, allowing the royal cabin to be uninterrupted

Her building began in May 1814 and she was launched on 9 August 1817. Her hull cost £14,921 for materials and £5187 for workmanship, but her masts and yards were even cheaper than for the *Royal Sovereign* – £6 for materials and £10 for workmanship. Materials for her rigging and stores cost £1107, but only £13 was needed for workmanship.[11] She was named after George, the Prince Regent, and not his father – she was consistently referred to as 'his' yacht. Unlike other yachts, she was based at Portsmouth rather than Deptford, closer to the prince's favourite resort of Brighton, and he would soon discover the delights of sailing her in the Solent.

～ The minor yachts ～

In 1800 it was proposed to build two new yachts in place of the old *Katherine* and *William and Mary* and a draft was prepared for ships 75ft long and 23ft broad, of 197 tons. In the event only the *William and Mary* was built, as slowly as any yacht of the period, and finally launched in 1807. Her upper deck had two large cabins aft at different levels as usual, but the plans give no indication of allocation to royalty and courtiers. In fact, she would be used by the Lord Lieutenant of Ireland.

The *Prince Regent* was designed as an exercise by the students of the Royal Naval College in Portsmouth in 1814/15,

The *Royal George* at Portsmouth in 1820, with the new King George IV on board – he is probably the figure wearing the Garter star to the left of the group of three. The signal tower is seen in the background. (Royal Collection Trust / © Her Majesty Queen Elizabeth II 2022, 404820)

perhaps because the imminent end of the wars with France left little chance to design new warships. There was nothing radical about her design, she was similar to the *Royal Sovereign* in dimensions and had the common flat-sided midship section with a 'V' bottom below. She used the now standard layout of two decks with stateroom, royal bedroom, another stateroom, 'noblemen's dining room', and crew's quarters, with officers' accommodation below. Again she was built slowly; by the time she was launched on 30 May 1820, George III had died and the former Prince Regent had been King George IV for more than four months. There is no sign that she was used much by the royal family. In 1836 it was proposed to present her to the Iman of Muscat, but that did not happen. She was broken up in 1847.

The *Princess Augusta* was sold to a Mr Ledger in August 1818 for £650, and in the following year he advertised the sale of:

The hull of a remarkably fine ship (late the *Princess Augusta* yacht), burthen 218 tons, built in his Majesty's yard at Deptford; has been thoroughly examined and raised upon in the most judicious and substantial manner; is generally copper fastened, every beam-end double kneed; her topsides and upper deck new, has new wales and a bust head and sham galleries, and now rendered capable of carrying upwards of 300 tons: well adapted for the coffee or Mediterranean trade. Lying in the City Canal.[12]

The old *Royal Charlotte* of 1749 was taken to pieces at Woolwich in July 1820 and her successor would represent the changes in the concept of royal yachts over seventy years. The draft of the new *Royal Charlotte* was approved by the Admiralty in December 1818. It had been drawn by the Joint Surveyors of the Navy, Sir Henry Peake and Sir Robert Seppings. The latter was a great innovator in ships' structure, and that is perhaps reflected in the plans he produced. The most striking feature was the tentative use of iron knees to support the beams of the lower deck: 'One beam next afore and one beam next abaft the mainmast, and one beam next afore and two beams next abaft the mainmast, and one beam next abaft the mizzenmast to be kneed at each end with an iron knee'. This was not repeated on the upper deck where they would be more visible to the royal passengers. Seppings was greatly concerned with the lead ballast, which was apparently shaped to fit in particular spaces, and he wrote, 'From station H to 15, on the square body to fill in between the timbers between from the line at A downwards to the head of the crosspiece B, with lead, and to be well caulked before the plank of the bottom brought on.' This was parallel to his policy with major ships, of filling the spaces between the lower frames to prevent movement; but he did not go so far as to deploy his more famous system of diagonal framing, which would have taken up space in the lower part of the hull. He produced a framing plan on which he wrote, 'All the timbers to be formed into frames', but it is not clear what he meant by that. The design and layout of the hull were less innovative, with the typical flat-sided and 'V'-shaped midship section and the usual layout of cabins on the upper deck. She was used by the Lord Lieutenant of Ireland.

By the 1820s only the *Royal Sovereign* and *Royal George* were fully-fledged royal yachts, with King George IV favouring the latter. They were both square-rigged and were much larger than most of their predecessors at 278 and 330 tons. They had finally left behind the legacy of the Stuart royal yachts,

The yacht *Royal Sovereign* off Weymouth, hove-to with the fore and main sails balancing each other, perhaps to allow guests to embark or disembark. The deck is crowded with passengers. (© National Maritime Museum, Greenwich, London, BHC3613)

6
The End of the Wars

The restoration of the Bourbons

In the spring of 1814, after more than two decades of almost continuous warfare, Napoleon Bonaparte's great empire was beginning to crumble. His invasion of Russia in 1812 had led to disaster and caused a powerful alliance against him, while the Duke of Wellington was leading an army which crossed the Pyrenees from Spain and advanced northwards. Napoleon abdicated on 11 April. Yet again the royal yachts took part in a regime change, albeit it in France rather than Britain. Louis, Comte de Provence, the younger brother of the executed Louis XVI, had been living quietly in England, but now the rules of succession, and the support of the allied armies, made him King Louis XVIII of France. He proceeded to Dover among great popular celebration. On 24 April it was reported:

The *Royal Sovereign*, the *Queen Charlotte* [sic], and other yachts, with the King of France and suite, Duke of Clarence, Duchess D'Angoulême, Spanish and Portuguese ambassadors … and many other distinguished personages, sailed from Dover Harbour this afternoon, at one o'clock, the guns all round the coast firing a royal salute. His Royal Highness the Prince Regent, who arrived yesterday afternoon, walked to the North Pier head, and joined most heartily in the cheers which resounded from all quarters at this happy event.[1]

The press compared it with the landing of Charles II in 1660, though they did not mention the arrival of George I a century earlier.

The yacht *Royal Sovereign*, flying the white Bourbon flag, conveying King Louis XVIII to France in 1814. (© National Maritime Museum, Greenwich, London, BHC3612)

The passage of two hours and ten minutes was said to be the fastest ever remembered.

> On arriving off the French coast, the royal yacht hove-to, when his Royal Highness the Duke of Clarence, in the *Jason* frigate, passed her, fired a royal salute, and then manned his yards, gave three cheers, and bore away; every ship of the fleet passed the royal yacht, saluted, and cheered. The *Royal Sovereign* yacht then approached the harbour of Calais, and was received by a roar of cannon which lasted upwards of two hours – The French coast, from Calais to Boulogne, appeared one entire blaze.[2]

Clarence had served in the navy including a period under Nelson in the West Indies in the 1780s, and reached the rank of Admiral of the Fleet in 1811. He returned to the Downs but Captain Sir Charles de Poer Beresford wrote from the *Royal Sovereign*:

> I have the honour and the satisfaction of acquainting your Royal Highness that having landed his Most Christian Majesty the King of France, his suite, and everything belonging to them in safety, I was in perfect readiness to have quitted this harbour with the whole of the vessels under my orders this morning on my return to the Downs, but unfortunately the wind continues so much from the northward that I fear it will be found impracticable even for the smaller vessels to get out of the harbour this day. I have therefore dispatched Lit Phillips in Vice-Admiral Foley's barge with this letter for your Royal Highness's information.
>
> From the nature of the tides at this time, the harbour master and pilots think it will be impossible for this ship and the *Royal Charlotte* to get over the bar for seven days, should we fail in our attempt this afternoon tide.[3]

They finally arrived off Deal on 2 May.

The arrival of the Allied sovereigns

Meanwhile, the Duke of Clarence was given another role. The King of Prussia, the Tsar of Russia and Marshal Blücher of Prussia headed the allied armies at the capitulation of Paris and it was arranged for them to cross the Channel to visit England. The role of the yachts was secondary. The sovereigns themselves were to go on board the 98-gun *Impregnable* with some ceremony, with parts of their retinues, while the rest boarded the yachts.

> Captain Eyles of the *Royal Charlotte* being the senior officer in the command of a yacht will accompany the suite of the Emperor of Russia in the boat destined for his standard, and Sir Edward Berry of the *Royal Sovereign* that of the King of Prussia, and as soon as they perceive the standards of Russia and Prussia displayed at the fore and mizzen topgallant mastheads of the ship, those previously displayed in the boats will be hauled in.

The remainders of the royal retinues, according to Clarence, were 'to be taken to the *Royal Sovereign* and *Royal Charlotte* yachts'.

Clarence was outraged to be told to haul down his flag on completion of the task and wrote on 9 June:

> I am this instant favoured with your lordship's letter of the 7th June and must confess that it has given me considerable pain, after having been selected to carry the King of France and there receiving universal approbation, and now being equally called upon to bring to this country the allied sovereigns, which service I will affirm has been most completely executed.[4]

He sent his servant to confirm that the order was valid, and in the meantime, in an act of defiant insubordination, he kept his flag flying until his return. It was not the last time the Duke of Clarence defied authority, but for the moment it was accepted by the authorities.

After numerous visits and ceremonies in London, Oxford and elsewhere, the sovereigns arrived at Portsmouth and made their first visit to the fleet on 23 June, being rowed out to the ships in the Solent in a procession of barges. They boarded Clarence's flagship *Impregnable* where the Tsar sampled the rum 'which you call grog' and proclaimed, 'I think it very good.' Next day the Tsar went out to the *Impregnable* with Clarence, while the Prince Regent and King of Prussia were rowed out to the *Royal Sovereign* yacht. She set sail, followed by a fleet of warships in a line seven or eight miles long off the Isle of Wight. They passed St Helens in a brisk northeasterly wind. 'The *Royal Sovereign* yacht led the van. The yachts and barges of the Admiralty, the Naval Commissioners, the Ordnance and other public offices, and a great number of private yachts, and above 200 vessels of all descriptions sailed out, keeping various distances from the fleet.' At five in the afternoon, with the leading ships about twelve miles from Portsmouth, the ships of the line were signalled to heave to and the Prince Regent and King of Prussia went on board the *Impregnable*. The winds were not helpful on the way back but 'the amazing accuracy of the naval movements, was of the most beautiful and of the grandest kind imaginable' – which must have impressed the visitors with the scale and competence of British sea power, if they were not too involved with the entertainment offered below decks.

The sovereigns left Portsmouth on the 25th and proceeded towards Dover. Two days later the King of Prussia sailed for Calais in the *Nymph* frigate. At seven in the evening the Tsar boarded the *Royal Charlotte* yacht, which had been drawn up against a wharf in the harbour so he only had to step on board. He was greeted by Clarence, Admiral Foley, who had led Nelson's attack at the Battle of the Nile in 1798, and other officers. As they sailed, the Tsar gained popularity by bowing to the assembled multitudes on pier heads and quays. They were bound for Ostend but the wind was unfavourable and they anchored in Calais Roads overnight. The Tsar landed at Calais at seven next morning.[5]

It was not the end of the wars. Conflict with America continued into 1815 but, much more seriously, Napoleon escaped from exile in Elba and resumed command of his army His defeat by the Duke of Wellington and Marshal Blücher at Waterloo was 'a near run thing' but it secured the peace of Europe.

The Prince Regent learns to sail

Britain remained a troubled country in the years after 1815. From 1812 the Luddites had smashed factory machinery which they believed would put them out of work. Notables like Byron, Shelley and Cobbett wrote against the Prince of Wales, now the Prince Regent during the King's illness. He was jeered during his public appearances, and his estranged and wayward wife was far more popular than he was, despite her many faults.

The prince began to take more interest in the yachts, perhaps as an escape from his troubles. 'His' yacht the *Royal George* was launched in August 1817 and was ready to join him off his favourite resort of Brighton in September. According to the local newspaper:

> On Monday, his Royal Highness, with the noblemen and gentlemen of his suite, embarked in the *Royal George* yacht, which, with the *Inconstant* frigate, had brought into the roadstead the preceding day. His Royal Highness put off in a boat facing the south end of the Steyne, and the instant he was a float, royal salutes were discharged. … The royal yacht now stood out to the SW with her armed companions, followed for some distance by the numerous pleasure boats, filled with gay company, and every species of floating craft that could safely be used in such pleasurable excursion. The day was lovely, scarcely a cloud was visible, and a light sailing breeze gave a gentle ripple to the surface of the ocean. His Royal Highness continued at sea for nearly ten hours. He was eventually landed from the state barge of the *Royal George,* at the embarking point of the morning, towards nine PM.[6]

The prince had 'enjoyed so much pleasure' that he intended to repeat the exercise the following day, but was prevented by the arrival of a messenger on government business. By Wednesday more warships had joined and a mock battle was staged with the *Inconstant* and *Grecian* 'defending' the yacht against the *Tigris* and *Rosario*.

> Some skilful manoeuvres now followed, and may tremendous broadsides sent their rattling reports to land. The escape of the yacht effected, the conflict between the frigates became more severe. … The battle over, which ended in the retreat of the enemy, the *Royal George* stood in, and the whole returned to the road-stead about two PM, without bringing up, but commencing a progress thence in the SW soon after.[7]

The prince decided to stay on board and some of his belongings were sent out from the Pavilion. On Thursday they set off early despite heavy rain, but they were unable to see the French coast before returning to Brighton early in the afternoon. Next day they arrived off St Valery, much to the surprise of the French. They signalled for a pilot, who turned out to be 'the very man … who had piloted the King of France into Calais' and who 'well explained the whole of the coast.' They stayed off the coast for some time with the prince remaining on deck, then returned to Brighton at one on Saturday afternoon. After that the prince was obliged to return to London on business, but '[h]e was gratified beyond description, and enjoyed the highest state of health and spirits during the whole of the excursion.' His new-found love of sailing was a surprise to many and it was reported that: 'A noble earl … took 50 guineas lately from a naval baronet, to return one guinea for every league the Prince Regent should sail on salt water within a period of two years – his Lordship would be very glad to hedge his bet, now the Prince Regent has braved the ocean.'[8] But the prince had a personal tragedy in November when his rebellious but beloved daughter Charlotte died in childbirth.

The Isle of Wight regatta was held during July 1818 and interest 'was not a little increased by the presence of the Prince Regent's yacht (the *Royal George*) which came down for her anchorage at Spithead, having on board Earl Spencer, Mr Tier and a large party of distinguished persons'. But the prince was

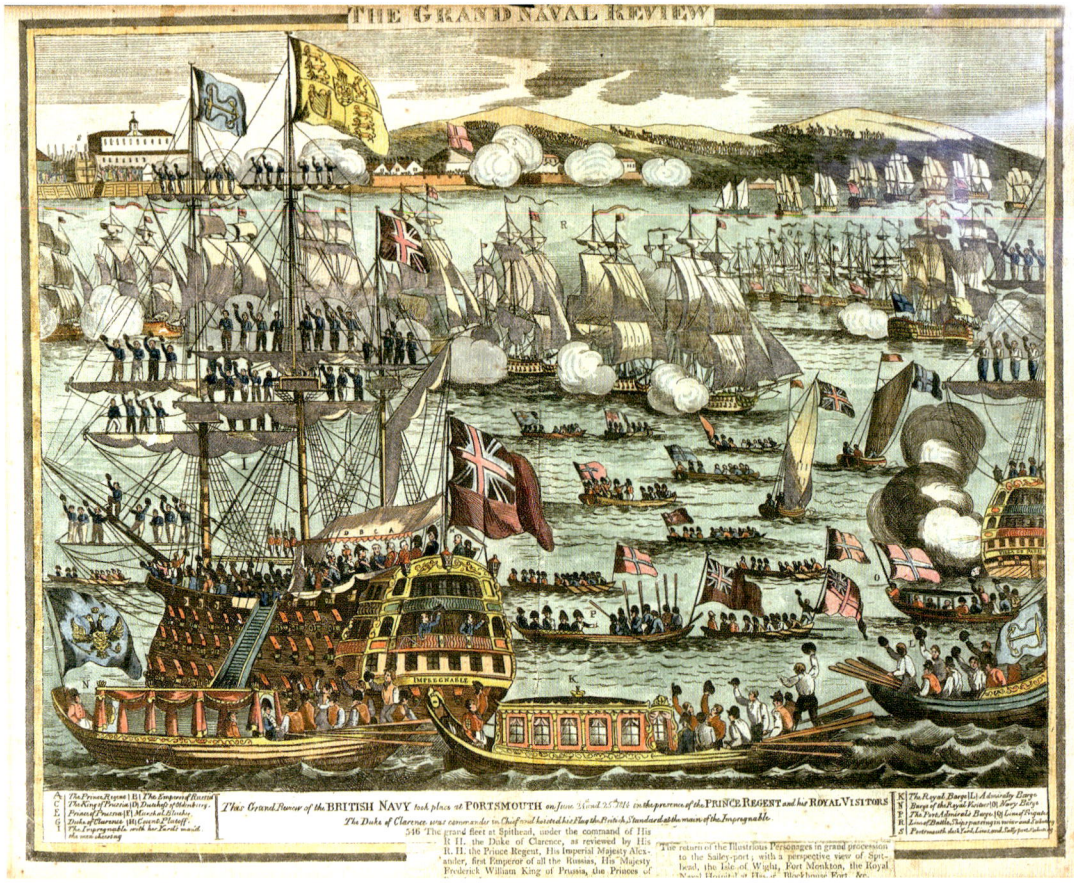

A cartoon-like representation of the naval review of the Allied sovereigns in 1814. Sailors man the yards of the flagship *Impregnable*. (© National Maritime Museum, Greenwich, London, PAF4791)

unable to undertake an excursion that year and the yacht was laid up at Portsmouth in September.

The prince made up for it in the following year. He boarded the *Royal George* at Brighton on 9 August and sailed to Portsmouth with the *Hyperion* frigate, the *Hind* and *Brisk* of 20 guns each, the *Chameleon* of 10 and two revenue cutters. They went to Cowes Road but did not land; instead the local dignitaries were entertained on board. They sailed up and down the Solent on the 11th and in the evening the prince went ashore dressed in an admiral's uniform for dinner at Admiralty House, returning to the yacht in the evening through the Sally Port in the sea wall. The 12th was noisy, even by the usual standards of royal visits. 'We had noise and smoke here all day long yesterday: the guns from the ships saluted the new flags in the morning, on account of the promotions of flag officers; and at noon all the ships and forts about the neighbourhood saluted the prince on account of his birthday.'[9] They sailed to Cowes Road on the 12th and this time the prince landed to dine with the Marquess of Anglesey. On the 13th they witnessed a regatta of about thirty vessels of 25 to 40 tons between the Bramble Bank off Cowes and the buoy where the great flagship *Royal George* had tragically sunk in 1782. Already it was being noted, 'The Prince Regent's delight in yacht-sailing increases daily'.[10] There was another regatta off Hyde in Southampton Water the next day and then the yacht returned to Cowes. They began to settle into a pattern and it was reported on the 25th, 'Not a day has passed since Friday, without the yacht being under way at eight AM, sometimes bending her course to the Needles, at others to St Helens or the coast of Sussex; and occasionally, when blowing very strong (which it has done for three days this week), to Southampton River, and parts within the Wight.' They returned to Cowes between five and six every day, the prince entertained a party for dinner and retired to his sleeping cabin at ten. It was noted, 'His Royal Highness, from thus constantly cruising, and sleeping on board his yacht, which he has done nearly 25 nights, has become an excellent sailor, bidding defiance to the discomposure usually felt by young yachtsmen, and sometimes by old nauticals, when Neptune chooses to be unpleasant.'[11] He bade farewell to the officers of the yacht on the 28th.

During the voyage the most dramatic of all the events of the post-war years had occurred. A crowd assembled in St Peter's Fields, Manchester, was attacked by yeomanry cavalry, leading to the death of eleven people, an event which entered popular history as 'the Massacre of Peterloo'. George's support for the Manchester magistrates augmented his reputation as a tyrant. His situation had changed drastically by the next sailing season, for the aged and ailing George III died on 29 January 1820 and his 57-year-old son ended his long wait to become King George IV.

Part V

Uniting the Kingdom

1

George IV in Ireland

King George's first problem in his new reign was with his long-estranged wife, Queen Caroline, who had arrived in the *Princess Augusta* in 1795. Her scandalous behaviour during her continental exile did not prevent her coming back to reclaim her rights as queen. The King's popularity dipped, but was restored by the time of his coronation on 19 July 1821, from which the Queen was turned away. The King immediately began to plan an expedition to Ireland, a project he had harboured for some time. He had been refused the office of viceroy some years earlier, his illegal marriage to the Roman Catholic Marie Fitzherbert had raised hopes in the country, and in 1797 he had planned to travel there to do 'justice to the ill-used Irish'.[1]

George IV embarking at Kingstown in September 1821. (Royal Collection Trust / © Her Majesty Queen Elizabeth II 2022, 750803)

Ireland

The King arrived at Landport Gate, Portsmouth, at five in the afternoon of Tuesday, 1 August 1821 and was presented the keys to the city by the lieutenant governor. He handed them back symbolically and the gates were thrown open. Avoiding the Royal Dockyard, he proceeded through the High Street of the old town, lined with troops, to the Sally Port near the harbour entrance. He boarded the royal barge, steered by Commodore Sir Charles Paget of the *Royal George*. The yacht was waiting for the King, having been moved in from Spithead due to bad weather. He dined on board at 8.30, while passengers in boats tried to get a glimpse of him through the windows – to gratify them he appeared on deck occasionally to greet them 'with that gracious condescension which distinguished him'.[2] The visit coincided with the arrival of the *Camel* storeship bringing the suite of the late Emperor Napoleon Bonaparte from St Helena, where he had died three months earlier.

The yacht got under way at ten next morning, being saluted by the Platform Battery and Fort Monckton. It was

followed by a squadron including the *Royal Sovereign* and *Emerald* yachts, the aptly named frigate *Liffey*, another frigate, four smaller warships and private yachts belonging to Lord Craven, Mr Pelham and the Marquess of Anglesey, who as Earl of Uxbridge had famously lost his leg at Waterloo and was one of the first people to be fitted with a fully articulated prosthetic. They anchored off Cowes and the King entertained a party on board that evening. They sailed on Friday morning. The largest ships, the frigates *Liffey* and *Active*, went eastwards to round Bembridge on the Isle of Wight, the others took the shorter but narrower route through the Needles Channel. The wind seemed favourable at first and according to the press on the following Monday there was 'little doubt that the royal squadron have, ere this, got round the Land's End into St George's Channel, in which case, His Majesty will, in all probability, reach Dublin either this evening or tomorrow.'

That was not to be. The wind shifted and at five in the afternoon of the 2nd the squadron anchored in Swanage Bay off Poole, and 'every one was anxious to pay a tribute of loyalty to his Sovereign, and every boat in the neighbourhood was immediately filled with spectators, all eager to have a view of their beloved sovereign.' The King stayed on deck for two hours, 'acknowledging the repeated salutations of his subjects'. The yacht was illuminated at night, 'which had a most pleasing effect', and they sailed at five next morning – but progress was slow and they were still in sight at noon, until the wind increased. The *Liffey* took the opportunity to punish four of her men with twelve to thirty-six lashes each. On the 4th the ships were becalmed and anchored off Weymouth on a site which would been familiar to the King's late father. There was thick fog during the night, but a fine breeze sprang up later in the morning. The yachts took the narrow passage between Portland Bill and the dangerous tidal race, affording a fine view to thousands of cheering inhabitants assembled on the cliffs under a Royal Standard, with a band playing *God Save the King*. Two locally based naval lieutenants went out in a boat and tried to deliver a 'loyal and congratulatory address' from the inhabitants, but this time the wind was strong and they had to deliver it to the *Liffey* instead. The squadron was soon out of sight.

At six in the morning of the 5th the squadron was seen from Penzance, which gave time for the populace to assemble at Land's End. Being a Sunday, the fishing boats were unable to launch in time, but the King and his consorts were seen three leagues northwest from the headland. The *Liffey* recorded a slow passage in light breezes up St George's Channel, during which she sent a boat over to supply the *Royal George* with water, as she did several times during the voyage, for the storage capacity of yachts was small. Holyhead was in sight by noon on the 6th and the frigate was making 'all possible sail'. They manned the yards as the King disembarked, and again when he was hoisted back on board.[3]

It was planned that the King would be entertained by the Marquess of Anglesey in his seat at Plas Newydd in the Menai Strait, then the yachts would show themselves only briefly off the harbour at Holyhead. Again the weather intervened and the *Royal George* entered Holyhead. The King landed to an improvised salute by two guns placed before the Customs House and boarded Anglesey's coach, which reportedly formed part of a procession at least a mile long, to be taken to Plas Newydd.

There was a complication, as it was reported that the Queen's health was deteriorating after her humiliation at the coronation. Despite the extremely bad relations with her, it would not be seemly for the King to ignore the issue. He accepted that 'it was utterly impossible for me, under any circumstances, not to proceed now to Ireland', but adopted 'the best medium line' that he should plan for a low-key entry at Howth to the north of Dublin rather than Dun Laoghaire (Dunleary) to the south as planned. And the idea of using a novel form of transport was raised: 'it was now determined that, either in the course of the day, as soon as possible as the wind and weather will permit (but which at present does not appear very encouraging) we are to set sail either in the yacht alone, or by steam to Ireland'.

At last, as it was reported:

> When the King's messenger arrived announcing the death of the Queen, he proceeded instantly to the King's yacht. Lord Londonderry happened to be on the deck; and when the messenger appeared it was notified to his Lordship. He instantly proceeded towards the messenger, who handed him a red morocco box, containing the dispatches, and at the same time said to him something in a whisper. His Lordship placed the box under his arm, and hurried down to the King, who was sitting in his cabin.

The King ordered mourning by lowering the ships' flags, but not the firing of minute guns, which was usual on such occasions.

It was nearly ten years since Henry Bell had made a success of his steamship *Comet* on the River Clyde, and since then the new form of propulsion had spread throughout much of the British Isles and Europe. It was still constrained by simple engines, low-pressure boilers and paddlewheels, so it was mainly confined to rivers, estuaries, coastal voyages and sheltered waters, which included the Irish Sea. Earlier that year the Post Office had been persuaded to accept its first two steamships for the route between Holyhead and Howth, near Dublin. The *Lightning* was built under naval supervision by Elias Evans at Rotherhithe, where some of Charles's yachts had been built. She had Sir Robert Seppings's diagonal framing, which made her hull much stronger and able to support the weight of two Boulton and Watt side-lever engines of 80hp each, with an exceptionally tall funnel, even for the times. She was launched in May 1821 and made a rather difficult to passage to Holyhead in bad weather, arriving on 29 May. During the rest of the year she made 140 crossings, an average of a day and a half for each without allowing for time in port. She was commanded by John McGregor Skinner, a popular local figure, who had been born in New Jersey in 1861 but joined the Royal Navy as a midshipman at the start of the American Revolution and, like Nelson, lost an arm and the sight of an eye. Over the last few days the *Lightning* had been involved in carrying officials, including the King's secretary, Sir Benjamin Bloomfield, and the Home Secretary, Lord Sidmouth, between there and Dublin in order to confirm the details of the royal visit and be ready to convey them to the King.

At eight in the morning of the 12th the *Lightning* hauled alongside the *Royal George* and the King, his suite and his luggage were transferred. Accommodation was spartan by royal standards – a ladies' cabin with six two-tier bunks and a gentlemen's with fourteen – but the voyage was short and the King made the best of it. According to a courtier, 'The passage to Dublin was occupied in eating goose pie and drinking whiskey, in which his Majesty partook most abundantly, singing many joyous songs, and being in a state, on his arrival to double in sight even the numbers of his gracious subjects assembled on the pier to receive him.'[4] The numbers were not so high in any case, perhaps five hundred people who 'had been attracted to the spot in the vague hope of seeing his Majesty arrive', while tens of thousands were at Dun Laoghaire. The King was said to be 'obviously much fatigued', which is not surprising after such a night, and descended a carpeted ladder. He was quickly whisked away to the viceregal lodge in Phoenix Park: 'not a single red coat, not a soldier nor a police officer appeared at the landing station, or escorted his Majesty on the way to the park'.

The King embarked on a round of levees, balls, concerts and banquets. Whenever he appeared in the streets he was greeted by large and enthusiastic crowds, both Protestant and Catholic. He made generous gifts to the common people he met. He gained credibility by meeting Daniel O'Connell, the leader of Catholic opinion and a very effective critic of government policy. The visit was judged a great success at the time, but soon doubts began to creep in. In October Charles Wynn wrote to the Duke of Buckingham, 'Meanwhile the state of Ireland seems to show that the blessed conciliation effected by H M's visit is confined to those districts which have been illuminated by his countenance.' And his populist style did not please everyone:

> Drinking toasts, shaking people by the hand, and calling them Jack and Tom, gets more applause at the moment, but fails entirely in the long run. He seems to have behaved not like a sovereign coming in pomp and state to visit a part of his dominions, but like a popular candidate come down on an electioneering trip.

A tempest

It was decided to return in the *Royal George*, departing from Dun Laoghaire, which was renamed Kingstown in honour of the occasion. After twice having to return, the yacht set off on Friday after 'a change of wind in our favour', according to the King. Winds were light to start with and the *Liffey* set studding sails. They headed south but within thirty miles of Land's End the weather turned much worse.

> About two or three in the evening the wind shifted immediately in our teeth; a violent hurricane and tempest suddenly arose; the most dreadful possible of nights and scenes ensued, the sea breaking everywhere in the ship. We lost the tiller, and the vessel was some minutes down on her beam ends; and nothing, I believe, but the undaunted presence of mind, perseverance and courage of Paget preserved us from a watery grave. The oldest and most experienced of our sailors were petrified and paralysed; you may judge somewhat, then of what was the state of most of the passengers; every one almost flew up in their shirts upon deck in terrors that are not to be described.

During 'a most formidable tempest … such as has been hardly known by the most veteran sailor … most of our fleet were separated, except the *Royal Sovereign* yacht, the *Liffey* frigate, and ourselves. Most of our crew and company were deadly sick, but the very worst of all as my poor self.' They reached the large natural harbour of Milford Haven in southwest Wales and the King wrote, 'now I am, for the first time, since we are again at anchor in smooth water, risen from my bed'. He was 'completely shattered and torn to pieces by the effects and sickness of an eight-and-forty hours tempest'.

An overland trip was considered, but it seemed even worse in view of 'the very mountainous and bad state of the roads through this part of South Wales, the scarcity of horses, the dreadful length of the stages, and, after all, the formidable length of the journey itself to London'. The King was resolved 'to wait for the first steady and favourable wind … that will carry us to Land's End in about eight hours; after which we may make Portsmouth at the very latest twelve hours afterwards, let the wind be then almost what it may.'[5]

But the weather did not get better and government business was increasingly pressing. At 5am on the 12th the

King entered a coach to begin the long journey. He was welcomed in Carmarthen and arrived at Brecon that evening. He passed through Abergavenny, Cheltenham and Henley-on-Thames and arrived in London on Saturday 15th. He had been away for forty-seven days, of which twenty-four were spent travelling by sea and land.

Hanover

The King's urge to travel had not been dampened. By 5 September while still in Ireland, he was planning to leave for Hanover on the 16th or 17th, but the delays to his return postponed that. By the 16th he had a very grand plan.

> Supposing he sets off the 24th or 26th; a fortnight to Hanover, as he goes through the Low Countries and visits the King of the Netherlands; this would make it the 10th October. A month there, 10th November. A fortnight's journey to stay at Vienna, 24th November. A fortnight more from Vienna to Hamburg, Wurtemberg, and to Paris, 10th December. Four days at Paris, about the 15th or 16th December return. And all this with the present state of the country. I do not think it possible for him to be allowed to do it.[6]

This was much reduced, but still included a trip to his ancestral realm of Hanover. The King chose a short sea crossing, starting from Ramsgate. On 25 September the royal physician Sir William Knighton wrote, 'Here we are safe in Calais, thank God, all well – a rough passage towards the close.'[7] He travelled via Brussels, Namur, Dusseldorf and Osnabruck. At the field of Waterloo he visited the willow tree under which was buried the leg of Lord Anglesey. By 12 October, 'The yachts are ordered to be off Calais the beginning of next month, and the King is under engagement to be back by the 9th.' The King arrived there on the 2nd, dined and went on board the *Emerald* in the harbour, which was supported by the *Dasher* and *Venus* steamships. He was rowed out to the *Royal George* anchored offshore and was in the Downs next morning awaiting a tide to take them into Ramsgate, the King's choice of landing place. He stepped ashore at the same spot from which he had left and went to London via Canterbury and Chatham, where the troops assembled to salute him. After such successes, the King began to consider a voyage to another of his dominions in the following year.

2
To Scotland

On 5 June 1822 Charles Wynn reported: 'The King now again proposes going to Scotland. The visits are to be to the Duke of Athol, Duke of Montrose, Lord Mansfield, and Lord Hopetoun; perhaps Lord Breadalbane, but not to Gordon Castel or Inveraray – the first on account of distance, the later of the Duke's absence.' Two weeks later the scope of the visit was more limited – 'The King is to be attended in Scotland by [Sir Robert] Peel and Lord Melville, but not to pay any visits; he is to be quartered at Dalkeith and his suite in Holyrood House' – for the ancient palace at the end of Edinburgh's Royal Mile was not in good condition. By the 23rd he seems to have turned against the journey: 'the K–, I should fear, is not quite in the good humour he was. He dislikes the journey to Scotland, and I have no idea why they plagued him to take it (which I believe to have been the case).'[1]

The Jacobite issue was dead as Scotland prospered due to links with the British Empire, and as yet there was no serious independence movement, but it was a troubled country in the early 1820s, perhaps even more so than England and Ireland. In 1820 the 'Radical War' centred in Glasgow and the Clyde led to the execution of three men and the transportation of many more. But the royal visit of 1822 would ignore the industrialised lowlands outside Edinburgh and the King would see nothing of the coal mines, cotton mills and ironworks which represented the future. Instead he would concentrate on a

George embarking at Greenwich in August 1822, with a great host of craft including the newfangled steamboats. (© National Maritime Museum, Greenwich, London, PAD6656)

TO SCOTLAND 111

largely mythical past with emphasis on the Highlands. The '45 rebellion was largely forgotten, kilted soldiers had greatly distinguished themselves in the wars with France, and the clan chiefs were now seen as the epitome of loyal subjects. It was with this in mind that Sir Walter Scott, at the peak of his fame as an epic poet (though he had not yet acknowledged his authorship of the highly popular Waverley novels) stage-managed the King's visit. He wrote to Macleod of Macleod:

> The King is coming after all. Arms and men are the best thing to show him. Do come and bring half-a-dozen or half-a-score of clansmen, so as to look like an island chief as you are. Highlanders are what he will like best to see, and the masquerade of the Celtic Society will not do without some of the real stuff ...[2]

And Scott would use the King to rehabilitate the kilt and tartan, banned since the '45 rebellion except for the army.

He planned a level of pageantry that would not have been possible in the past, when the arrival of any ship, even a royal yacht, was dependent on uncertain winds and the troops and dignitaries might have to stand by for weeks awaiting the royal arrival. But now steam power, though limited in its application to the open sea, was on hand. Steam was now prevalent in three areas of the United Kingdom – on the Clyde which might be seen as its birthplace, on the Thames where the boats could carry passengers downriver to resorts such as Margate and Ramsgate, and on the well-worn route of the Leith to London 'smack'. This provided the most efficient service between Edinburgh and the Thames, and Scott wrote to a friend that year, 'If you do not fear the sea the steamboat brings you close in to Edinburgh in sixty hours certain with as much ease as if you were in an easy chair, with all the convenience of public coaches'.[3] The newest ship on that trade was the *James Watt*, at 448 tons the largest steamship of the day when launched in 1821. She was about to leave Deptford on 8 August when orders arrived to transfer her passengers to the *Tourist* and fit her up to receive the King – for it was considered possible that he would go all or part of the way in a steamship. According to one report, 'comfort much more than magnificence had been studied'.[4] Moreover, the navy was not nearly as opposed to steam power as popular myth suggests. In 1821 it ordered the *Comet* (not to be confused with Bell's pioneering vessel) as a tug and ferry on the Thames. She was 115ft long and displaced 239 tons, with an 80hp side-lever engine. Both these ships would be available to assist in the King's voyage.

Even so, there was some uncertainty and the *Edinburgh Gazette* announced: 'When the probable time of his Majesty's arrival at Leith shall be known, public notice thereof shall be given', and the officials were to assemble.[5] And it was agreed that the King would remain on board the yacht for the remainder of the day of his arrival, to allow the parties to assemble. But it was decided not to put His Majesty in a steamboat again, with its soot and smoke and a highly dangerous boiler. He would go in the *Royal George*, but be towed when necessary by steamships. This facilitated a departure from the dignified surroundings of Greenwich, as they did not have to use the winds and tides to get down the winding river.

There were two false alarms at Greenwich as crowds gathered to watch the royal departure. Originally the King was expected to leave on 8 August, and '[t]housands of persons were seen advancing through the avenues leading to the centre of attraction ... Every house of public entertainment was thronged to excess, and the consumption of refreshments usually obtained in such places exceeded all former precedent.' But the departure was cancelled, and on the following day, Friday, 'a similar scene of disappointment was presented', leading to rumours that the trip was cancelled. There was, however, a firmer indication that day when the Lord Mayor of London, technically conservator of the River Thames as far as the Nore, received notice and began to prepare his barge at the Tower, ready to be towed down to Greenwich. The populace gathered again on Saturday morning, but even steam power had to take account of the tide and that was not suitable until the afternoon. Royal Marines from Woolwich arrived in the great square of the Naval Hospital, dignitaries came, including the Earl of St Vincent, and the Lord Mayor arrived, towed by the steamship *Royal Sovereign* (not to be confused with the royal yacht of the same name). Most of the City livery company barges were not ready due to short notice: only those of the Goldsmiths' and Skinners' companies were present.

The King's carriage left St James's Palace at quarter to three, preceded by wagons carrying luggage and servants. He arrived at Greenwich twenty minutes later to a general 'Huzza', rested briefly in the Hospital Governor's House then was rowed out to the *Royal George*. There was some delay in unmooring, but a hawser was attached to the *Comet* and the trip began. Dozens of boats followed, but soon the power of steam was demonstrated again. All except the steamboats had been left behind by the time they reached Woolwich. With a fine breeze, the yacht was able to unfurl her sails off the East India Dock. At Gravesend the massed population waited, with the Chatham division of marines. They had some indication of progress when they heard the salutes fired at Woolwich, and they could see the cloud of smoke put up by the steamboats. After Gravesend the wind came from astern and the yacht set studding sails to go even faster. The towing hawser was let go, but the two steamships stayed close to be of any assistance, which was required later. At nine in the evening they reached the Nore. The Lord Mayor considered that his duty was done as they approached the end of his jurisdiction and started back. Lieutenant John Boteler was in command of the *Seagull* cutter, having served on board the *Royal Sovereign* yacht for a time. He was ordered to the Nore to await the arrival of the royal flotilla. 'It was a dark and very still night, and the measured strokes of the paddle wheels was a new and most peculiar sound to us'[6] – though it would not have been if he had served in the Thames recently.

After a rest the squadron left the Nore at 4pm on Sunday 11th. As well as the *Royal George* and the two steamships, it consisted of the *Royal Sovereign* and *Prince Regent* yachts, *Phaeton* frigate, *Egeria* sloop, *Chameleon* and *Calliope* tenders and two other yachts. The *James Watt* was towing the *Royal George* and the *Comet* hauled the *Royal Sovereign*, until both steamships were ordered to attach themselves to the principal yacht. As they passed various ports and seaside towns, numerous local craft came out to greet them. They passed the convict ship *Czar* and the prisoners on board were allowed to cheer His Majesty. The wind was favourable during the night but the yacht was still using steam assistance, and 'made great way'. Next morning they sailed under the promontory of Flamborough Head with the usual cluster of boats, and the Mayor of Scarborough used a long pole to pass an address to the King on his birthday. The band on board the *James Watt* was transferred to the *Royal George* by boat and returned, after playing, with a hamper of wine. But the wind began to rise as they passed Tynemouth and there was a gale from the southwest. The *James Watt* cast off as the yacht could sail well enough on her own. By seven next morning they were off Eyemouth with the wind blowing down the Firth of Forth. The *James Watt* was attached again to counter this, but strong gusts coming off the hollows in the shore made conditions difficult. They anchored off Berwick-upon-Tweed and this allowed the *Royal Sovereign* to catch up. As usual, boats came out from the shore, and at Berwick there was 'great bustle in the town and neighbourhood' at the unexpected visit.[7]

Early in the morning of the 14th there was still uncertainty, and Commodore Paget went several miles to seawards in the *Comet* to take a look. It was decided to proceed and at seven they passed St Abb's Head, then the Bass Rock, which fired a salute. A message for London was put ashore at Dunbar, and at Aberlady Bay they were met by the *Queen Margaret* steam packet chartered by various ladies and gentlemen. They passed within forty or fifty yards and the passengers sang *God Save the King*.

The *Regent* revenue cutter was sent out from Leith to look for the squadron, and found the *Prince Regent*, *Phaeton*, *Egeria* and other vessels east of Inchkeith, then the two principal yachts. By this time the smoke of the steamers was discernible from the telegraph station at Leith, and soon the squadron was in sight. The royal standard was observed and a 21-gun salute was fired from all the ships at anchor in Leith Roads. Crowds hastened to the waterside to observe and cheer, but the yachts anchored and the King was not to land until noon next day. The main organiser of the event was rowed out and the King was pleased: 'Sir Walter Scott? The man in Scotland I most wish to see! Let him come up.' Sir William Knighton commented of the writer, 'He has no trace in his countenance of such superior genius and softness of mind as the beauty of his writing displays; but the moment he speaks, you discover a correctness of understanding and a display of intellect, marked by the utmost accuracy of thought.' Scott was allowed to keep the glass from which he drank cherry brandy with the King.[8] Overnight there was an immense bonfire at the top of Arthur's Seat. It was dry on the morning of the 15th when the King came on deck and 'beheld the Scottish capital, with its towers and palaces, basking in the rays of an autumnal sun'. A few minutes before twelve a gun from the *Royal George* announced that the King had entered the royal barge and it was rowed ashore, steered by Commodore Paget with a crew dressed in blue frocks and black velvet caps, and preceded by the barge of the admiral on the station. At twenty past twelve the King stepped ashore at Leith, the first monarch to visit Scotland since Charles II's disastrous campaign of 1651.

The Edinburgh that George was visiting was very different from the notoriously unsanitary and often fractious capital of the last century. It had already passed through much of its golden age, with luminaries such as David Hume in philosophy, Adam Smith in economics, Henry Raeburn (who would be knighted during the visit) in art, Joseph Black in science, Robert Adam in architecture and, of course, Sir Walter Scott in literature. The first phases of the New Town, planned in 1767, were complete and the elegant Georgian squares and terraces were a striking contrast to the ancient and often chaotic buildings on the hill above. The scheme was centred on George Street and many other names – Charlotte Square, Frederick Street, Queen Street, Princes Street and Hanover Street – proclaimed a post-Jacobite loyalty to the regime. Nevertheless, tradition demanded that the King spend much of his time in the Old Town, holding court in Holyrood Palace at one end of it, visiting the castle at the other and worshipping in St Giles's Cathedral in between.

Even Scott was amazed at the success of the visit, despite:

> the nature of the Scots people being stiff and haughty and distant … From the highest to the lowest they were anxious to know what was proper to be done and how to do it when they learned [it] as well for their own sakes as the King's. It was a very curious thing to see whole roads and streets lined with so many thousands of people who were (even the very meanest) all dressed in something like decent attire and each obviously considered himself as part of the spectacle and as having the national reputation dependent to a certain degree on his own behaviour.[9]

The King duly wore his kilt, though he insisted on flesh-coloured tights rather than bare legs. The visit had a profound effect on Scottish culture which lasts to the present day, shifting emphasis from the Lowlands to the Highlands and making tartan a universal national symbol rather than the dress of despised and lawless feudal clans.

The King's last trip on the 29th took him a little outside Edinburgh, to Hopetoun House, another symbol of the new Scotland in the heart of the old. It was an elegant mansion in contrast to the sturdy and defensible tower houses which had once dominated the countryside. Built on the profits of lead mining, the house had been designed by Sir William Bruce and the Adams family and consisted of a central building linked to wings by colonnades. The King departed from there to Port Edgar, less than three miles away, where a pier had been built in 1810. An ingenious ramp had been constructed to allow the royal barge to be boarded at any state of the tide, which was at half-ebb when the King embarked. The royal squadron was stationed in line with the pier, moored to a rope cable bridle. The barge set out 'with the velocity of an arrow set free from the shaft', surrounded as usual by numerous private boats. The *Royal George* was well equipped to feed His Majesty, with two bucks sent by the Marquess of Hopetoun, along with two boxes of fruit. They were towed by the *James Watt* and picked up a north-northwest wind off Leith so they set all sails. They were closer to Fife and out of sight of the southern shore but their presence was marked by a light from the masthead of one of the vessels. Off Whitby the favourable wind veered to the south, but the *James Watt* was following and took the *Royal George* in tow, as did the *Comet*. They passed Yarmouth on the 31st and at Greenwich it was anticipated that the yacht would arrive back on Sunday, 1 September, though the exact hour was not known. Nevertheless, numerous boats gathered, and the harbourmaster and Thames police had to make sure there was a clear space for the yachts to moor. The *Comet* went ahead down the river and at Greenwich a midshipman was sent ashore with a message that the King would arrive at four. Again, the smoke of the steamers was spotted and at a salute was fired at Woolwich. The yacht was off Greenwich Hospital by a quarter past four and the King was rowed ashore. The visit was another great success, but it began to raise questions about the future of purely sailing yachts.

The yacht *Royal George* passing Yarmouth on the King's return from Scotland in 1822. As usual the artist is coy about showing the steamship pulling the yacht. Fishing boats are carrying passengers out to greet her. (© National Maritime Museum, Greenwich, London, PX9799))

3
The Lord Admiral's Excursion

The King's brother, Prince William Henry, now Duke of Clarence, was heir to the throne after the death of Princess Charlotte in 1817. He had played a part in the restoration of Louis XVIII and the visit of the Allied sovereigns, and never totally lost interest in maritime affairs. During a rough passage across the North Sea to Antwerp in 1822 he stayed on deck all night, and in the morning he remarked casually to the officer of the watch, 'I think you want a pull at the main-top-gallant-brace.'[1]

In February 1827 George Canning became prime minister when Lord Liverpool was incapacitated. The long-standing First Lord of the Admiralty, Lord Melville, refused to serve under him and the ancient office of Lord High Admiral was revived and offered to Clarence. It had brought a great deal of power in the past, both ashore and afloat; Howard of Effingham had fought the Spanish Armada and James, Duke of York, the Dutch, but now the powers were severely restricted in the patent which appointed Clarence. In particular, he had no control over finance, and in other matters he was to rely on the guidance of his council, which included the forceful and experienced Admiral Sir George Cockburn. And appointments to the council were in the hands of the King, who would almost certainly rely on the advice of the prime minister. Canning probably thought of Clarence as a figurehead, but the duke had his own ideas on much-needed naval reform.

On 7 July 1827 Clarence boarded the *Royal Sovereign* at Deptford with Lord Errol and Messrs Sydney and Fitzclarence for an extended voyage round the southern naval bases, assisted by the steamships *Lightning*, *Comet* and *Meteor*. They passed through the Solent, but did not stop and headed on through the Needles Channel to Plymouth, where they reached the uncompleted breakwater at the entrance to the Sound on the 10th. After being saluted by nineteen guns each from the numerous forts, Clarence inspected the dockyard on the 11th and the Royal Marines on the 12th before dining ashore at the Naval Club. He inspected the various ships in the port on the 13th, hosted a levee on board the yacht on the 14th, then visited the new victualling yard after being joined by his wife, who travelled separately by carriage during the whole of the trip. He found time to hold a levee on board for the forgotten men of the navy, the half-pay officers, and for the forgotten ships which were laid up 'in ordinary' for any future war. He dined with the admiral and the commissioner, and on the 19th the yacht was hauled out to a mooring just inside the breakwater and dressed overall to watch a regatta. Finally, on the evening of the 20th she was towed out of the Sound by the *Lightning* and reached Falmouth the following day, where the duke visited the packet station.

They sailed in the evening of the 22nd to be saluted by the guns of Pendennis Castle and passed the Lizard and the Longships rocks off Land's End. The tow was cast off at 9.30 in the morning of the 23rd and the *Royal Sovereign* set studding sails and royals in light winds; but at six in the evening she had to be taken in tow again and in the process the *Lightning* 'got foul of our channels but did not do much injury.' Soon afterwards they reached the new dockyard at Pembroke, inside the great natural harbour of Milford Haven. On the 25th Her Royal Highness launched the new 84-gun ship, which was appropriately named the *Clarence*. After the usual round of visits and inspections the yacht sailed on the 27th. They were off the Longships rocks early in the morning of the 29th, under tow by the *Lightning*, when the port hawser was carried away and replaced, only to be followed by the starboard hawser, which was replaced in turn by one from the *Lightning*. But by 9.50 'the wind was too strong for the *Lightning* to tow us ahead. Cast off the tow ropes and made sail to the southward under close reefed topsails and courses' – a traditional method of coping with rough weather in a sailing ship. Towing was resumed in the evening and they passed Land's End and the Lizard.

They passed the Needles again on the 30th and this time they intended to stay. They were saluted by the Royal Yacht Squadron at Cowes, the first of several interactions with private yachts. The *Lightning* towed the yacht into Portsmouth Harbour and she was saluted by Nelson's old *Victory*, still afloat off the dockyard. Clarence inspected the dockyard and the ordinary and dined with the Mayor of Portsmouth. On Friday, 3 August the crew of the *Royal Sovereign* spent the day

The *Royal Sovereign* flying the Duke of Clarence's flag sails among the Russian squadron, with sailors manning the yards; by Henry Moses. (© National Maritime Museum, Greenwich, London, PAD8013)

preparing her quarterdeck for a dinner with seventy-six 'persons of distinction'. Saturday was spent inspecting troops and marines, and on Sunday the royal party attended divine service on board the *Victory*. They sailed up Southampton Water on the 6th, but only as far as Netley Abbey, where they turned and tacked down again. The 7th and 8th were spent with a Russian squadron of eight ships of the line, seven frigates and a sloop, which had arrived on the 3rd, supposedly on their way to the Mediterranean. The yacht was towed round the anchored squadron, which saluted Clarence. He entertained the Russian admiral and his officers on board the yacht, and in turn he visited the Russian flagship. In the afternoon of the 9th he went on board the 76-gun HMS *Warspite* to watch her crew exercise the great guns – something of an obsession with the duke. They began to head back soon afterwards with the usual combination of sailing and towing. They passed the well-known landmarks of Beachy Head and Dover Castle, had bad weather again in the Downs, where they had to lower the topgallant yards, but they were soon upriver at Deptford. Clarence was in no hurry to get ashore: he watched as the yacht dropped anchor in the river and was hauled alongside the *Bacchus* hulk. He finally landed at 9.30 in the evening of the 10th, after a pleasant voyage largely devoted to ceremonial duties.[2]

Canning died on 8 August while Clarence was at Portsmouth, and was replaced by the all-powerful and forceful Duke of Wellington, still basking in the glory of his famous victories over Napoleon's armies. Clarence soon exceeded government policy in his support for Sir Edward Codrington after he destroyed the Turkish fleet in support of Greek independence. Though bound to abide by the decisions of his council while on shore, he discovered that by a loophole in his patent he only needed the consent of one member while he was afloat. Thus on 10 July 1828 he issued an order dated on board the *Royal Sovereign* at Deptford (though Wellington suspected he was at home at Bushy Park at the time).[3] He called on members of a commission on naval gunnery to meet him in Portsmouth. Certainly, naval gunnery was ripe for reform, and the expense involved was small, but Cockburn complained, 'if your Royal Highness can, whilst so separated from the Council, issue orders concerning important questions of official regulations, or of public policy, or involving (in this instance) expense to an indefinite amount, the responsibility of the Council would be ideal rather than real'.

But Clarence had no intention of succumbing. He sailed in the *Royal Sovereign* and by the morning of the 11th he was off North Foreland. He wrote to Wellington suggesting that Cockburn be removed, which was not supported by the King. By the 12th the yacht was off Newhaven in Sussex, and in Portsmouth Harbour by the 13th. Clarence wrote to the King and he could not resist a dig at his rival: 'Sir George Cockburn *cannot* be the *most useful* and the most *important officer* in your Majesty's service, who *never* had the ships he commanded in *proper* fighting order'[4] – for Cockburn did not apply the rigid discipline favoured by the duke. Clarence returned to London but failed to resolve the matter.

Unrepentant, the duke rejoined the *Royal Sovereign* at Hamoaze, Plymouth, ostensibly following Sandwich's policy of dockyard visits. 'The eye of the First Lord of the Admiralty or the Lord High Admiral does infinite good, and the nation at large felt the advantage of Lord Sandwich annually visiting the arsenals from 1771 until the war with America and France.'[5] But on 1 August Wellington wrote to the King in some alarm: 'I am sorry to tell your Majesty that I received accounts this day that the Lord High Admiral sailed from Plymouth on the 30th July, in the *Royal Sovereign* yacht, which bore his flag as Lord High Admiral, with the squadron of ships and vessels as follows, viz – The *Britannia, Orestes, Pylades, Procris*.'

William had taken advantage of the fact that Sir Henry Blackwood – 'Nelson's watchdog' and one of the heroes of Trafalgar – had not yet arrived to take command of a squadron of exercise. William had no difficulty getting the support of the *Royal Sovereign*'s officers. The Honourable Captain Robert Cavendish Spencer had gained a reputation as an expert in gunnery, which perhaps attracted the attention of the duke – he was reputed to be the author of a catechism known as *The Ninety-nine Questions*. In any case he was appointed private secretary to the Lord High Admiral, as well as a Groom of the Bedchamber, and acting captain of the *Royal Sovereign* in June 1828. He was assisted by Lieutenant the Honourable Edward Gore, fourth son of the Earl of Arran and flag lieutenant to William as Lord High Admiral.[6] The master, G F Morice, was also a recent appointment; the surgeon and purser had been on board much longer, but they would take no part in executive decisions.

Off Land's End they practised gunnery, a particular interest of the duke, for despite some reverses against the Americans in the War of 1812, the navy had not moved far beyond Nelson's dictum that 'no captain can do very wrong if he places his ship alongside that of an enemy.' The yacht dropped a butt in the water and the *Britannia* fired ninety-eight shots at it, but it was hauled back on board undamaged after two hours. They practised again in the afternoon, but the cask remained intact. They exercised the sails in the afternoons, and there was a note of realism early on 2 August when squalls forced the reefing of the topsails. There was total contrast in the afternoon when the wind dropped and studding sails had to be set. The Russian squadron was sighted in the distance, but apparently not close enough to exchange salutes. They returned to Plymouth on the 3rd, and anchored.

Not surprisingly, the Duke of Wellington was furious. On the 9th, two days after Clarence's return to London, there was 'a most violent scene' between him and Wellington at the Admiralty, but the latter claimed, 'I completely calmed him, and got the better of him, and we parted the best friends.' Wellington wrote:

> When your Royal Highness embarked in a vessel bearing the military flag of the Lord High Admiral, went to sea from Plymouth with a squadron placed by his Majesty's command under the command of Sir Henry Blackwood … it was my duty, as his Majesty's minister, to submit to your Royal Highness the opinion I entertained … I entreat your Royal Highness to consider the terms of your patent, and the royal prerogative.

There was no alternative but resignation. According to Croker, the Admiralty Secretary:

> Clarence came to the Admiralty Council on the 14th and said that he looked upon himself as a military officer; that if he were a civil officer, like the First Lord of the Admiralty, he would have many observations to make on the cause of his resignation, but that, in his military character, he could only say that he had resigned, and would give no reason for it.[7]

He left office formally on 19 September and was succeeded by Lord Melville.

The *Royal Adelaide*

On the death of his brother, Clarence ascended to the throne as King William IV in 1830. There was a clean sweep at the Admiralty with the retirement of Melville, Cockburn and Croker, but William seems to have reached a higher level of maturity, even if his behaviour could sometimes be embarrassing. His reign saw the Great Reform Act of 1832, which removed some of the most blatant corruption from the electoral system, the abolition of slavery, and many other reforms. He tried to protect his niece Princess Victoria from her ambitious mother and abusive tutor, and in a sense his reign was a transition between the excesses of the Georges and a much more responsible and moral monarchy.

He seems to have had his fill of seafaring and the royal yachts were little used, though perhaps he had not tired of sailing. The *Royal Adelaide* of 1834 was named after William's beloved wife, and was intended for Virginia Water, a lake near Windsor Castle. She was designed by Sir William Symonds, who had strong ideas on naval architecture, attempting to design sailing warships on the lines of a sailing dinghy. She was a miniature version of his frigate *Pique* which was building at Plymouth, but at 50ft in length she was less than a third of her size. Like most ships of the period she had little sheer, that is her decks and sides were comparatively flat. She also had the defining feature of a Symonds design, a pronounced 'V' shape below, and more tumblehome than had been seen recently on yachts. She was made to look like a frigate with false gunports, but that was an illusion. She had only two decks, with a plain staircase leading to a lower deck at the level where the hold would have been on a full-sized ship. Her timbers were cut out at Sheerness Dockyard, she was assembled, then taken to pieces for transportation. She was to have a rather extreme sail plan with royals above the topgallants. These were retained (if not necessarily used), but the lengths of the yards were reduced so that the sails were less wide. She was to be armed with twenty-two captured French 1pdr guns in store at Woolwich, which were bore up to 1½lbs, and the royal carriage department made carriages for them.[8]

The parts were put onto a barge to be unloaded at Staines on the Thames near Windsor. There was trouble finding the screws which were apparently intended to hold her together, instead of the usual trenails and bolts – they were eventually sent on by road carriage. One leading man, two first-class and four second-class shipwrights were sent to assemble her, along with a caulker.[9] She was launched into Virginia Water with some fanfare in May 1834. The town of Windsor, it was reported, 'presented an unusual degree of bustle and gaiety' and the King had 'expressed his wish that all persons should be admitted on this side the Lake on foot'. As a result, 'there were several thousand persons present and among them a great number of the young gentlemen of Eton College'. The royal entourage arrived in ten carriages and embarked in the royal barge to the strains of *Rule Britannia*. The King named the ship and Mr Tinmouth, the naval officer in charge at Virginia Water, indicated that everything was ready, and 'the dogshores were knocked down and the vessel glided majestically into the Lake, everyone exclaiming, that a more beautiful launch could not have been witnessed', though she almost grounded in the shallow water. The royal party boarded the ship and the standard was hoisted to the cheers of the populace.[10] She would survive there until 1877, forty years after William's death, and perhaps played a part in introducing some of Queen Victoria's children to sailing. She fired salutes on royal birthdays, and in 1894 the Prince of Wales presented her guns to the Royal Yacht Squadron at Cowes.

Part VI

Building and Sailing

1

Captains and Crews

Charles's captains

Charles II's yacht captains were a mixed bunch, which perhaps reflects the navy as a whole. Some were 'tarpaulin' captains who had risen from the ranks or the merchant service. A few were well connected. William Saunderson was the grandson of one of Charles's courtiers; George Aylmer was the son of an Irish landowner and commanded the *Anne* for a time in the 1680s. Some are completely obscure, such as Peter Wotton, who is only known for being appointed to the *Cleveland* in 1678.

The monument to Sir Charles Molloy in the Church of St Peter and St Paul, Shadoxhurst, showing a bust of him in the centre, with a weeping cherub and various nautical objects around. (Alamy)

Some specialised in yachts, like John Clements, who successively commanded the *Cleveland, Charlotte* and *Henrietta*; William Fazeby of the *Katherine, Monmouth, Anne, Monmouth, Charles* and *Henrietta*; and Thomas Lovel of the *Henrietta* and *Katherine*. Andrew Cotton appears to have specialised in experimental vessels, including Sir William Petty's double-hulled sloop, or catamaran, in 1679, as well as the *Monmouth* and *Charlotte*. Command of a yacht might be merely an interlude to keep an officer in employment in peacetime, as with William Davie in the *Cleveland* and *Katherine* in the 1680s. He was one of the few to rise to high rank as vice admiral. The other exception was John Nevell, who also died as a vice admiral, but his connection with the yachts was very tenuous – he was promoted to captain of the *Anne* by Arthur Herbert 'in justice to Mr Nevills merit whose behaviour on many occasions has struck envy itself dumb', but immediately reverted to lieutenant when Herbert's authority to make such an appointment was challenged. Some were supporters of King James and their names disappear from the navy soon after the revolution – Sir William Saunderson, Anthony Crow who had commanded the *Kitchen* for eleven years, and Gabriel Millison, formerly of the *Katherine*. Others were supporters of the revolution and were promoted to ships of the line. George Aylmer was killed in command of the *Portland* at the battle of Bantry Bay in 1689.[1]

Captain Crow had a particularly chequered career and petitioned the government in 1696. He had:

> … served the crown nearly 40 years at sea, and in 1669 commanded the *Henrietta* yacht, in which he brought the King [presumably William of Orange] and Lord Ossory from Holland. He afterwards commanded the *Martin* [sic] yacht, and, after that, was made porter of Portsmouth yard, where he continued till the late King James dismissed him. He has ever since been pilot of the King's ships, but, as his eyesight is failing him, he is unfit for that service, as he is 63 years of age. He prays an order of superannuation in the same quality as when commander, for the subsistence of him and his family.[2]

Charles's captains used their privileges to the full, and sometimes stretched them. In 1667 the Treasurer of the Navy, Sir Thomas Allin, had to admit that 'none of the commanders of the yachts have heretofore done their duty as to the keeping of their muster books'. Captain Pinn, late commander of the *Cleveland* was mentioned in particular.[3] Men-of-war did not often visit foreign ports, but yachts did. By September 1675 the Lord Treasurer had received several complaints from the Customs 'that much merchant goods are brought in by the commanders of the King's yachts'. It was resolved that 'the respective commanders of the said yachts be severally writ to, strictly prohibiting them any such practice.'[4]

William Sanderson

William Sanderson (not to be confused with Sir William Saunderson, who apparently left the navy at the revolution) could trace his family back to 1387 and the days of Richard II. His grandfather had died in Royalist service in the Civil War, and his father and brother were both naval captains. William was in command of the *Henrietta* yacht at the breaking of the boom at Londonderry in 1689, a key moment in William's Irish campaign. He took command of the new *William and Mary* in 1694 and he transported the King to Holland several times, corresponding with the Navy Board about His Majesty's needs, such as a larger boat and a different size of glass in his cabin. In 1700 he went to the *Peregrine Galley*, which technically was not a royal yacht, though it served as such in practice in transporting the Duke of Marlborough. In 1714 he brought over the new King George I from Hanover. As a reward he was knighted at Gravesend, but he was replaced as captain by Galfridus Walpole, the brother of the future prime minister. The official justification was that Walpole had seen much action during the recent wars and had lost his arm at the Battle of Vado in 1711, whereas Sanderson had spent all his time in yachts. Sanderson complained about his dismissal, and claimed the right to keep some of the furniture taken from the yacht, including a bed which he wanted to pass on to his grandchildren.[5] Instead he was made Gentleman Usher (Black Rod) of the House of Commons.

Charles Molloy

The longest-serving captain of the royal yachts was Charles Molloy. His origins are obscure but he entered the navy before 1697 'when very young'. As a lieutenant he fought in the drawn Battle of Malaga in 1704 and became captain of the *William and Mary* in 1710. On 26 March 1726, 'This morning I attended the Admiralty Office and received from their Lordships a commission to command the *Mary* yacht being rebuilt and lately launched at Deptford. In the afternoon I went to board her and hoisted the pendant'.[6] He was something of a rough diamond. In 1737 Richard Stacey and the boatswain of the yard came on board the *Mary* in the wet dock at Deptford to ask Molloy to move it forward to allow lighters to unload. Molloy refused, according to Stacey, 'in such a manner as I have not been much used to' and 'added in a menacing manner, that he should take what he said, to whom, and that he would complain of him.' Stacey defined the relative powers of the dockyard officials and the captains, as he saw them.

When Molloy asserted 'that he was a commander there, and that the yacht should not be removed', Stacey replied that 'he might take his yacht and carry it where he pleased.'⁷ Despite this, Molloy was knighted by the King on board the *Mary* in 1743, and them promoted to the *Royal Caroline*. He took over her successor when she was launched in 1750.

It became increasingly common to appoint distinguished officers to command the yachts, partly as a retainer or a reward for services rendered. Peircy Brett was part of Anson's inner circle, having begun the famous circumnavigations as a lieutenant in the *Centurion* and rising to command the ship – Anson briefly resigned when the Admiralty refused to backdate his promotion. Brett commanded the *Lion* which crippled the French *Elisabeth* in 1745, severely damaging the prospects of the Jacobite Rebellion which followed. He became Anson's art adviser, producing many of the drawings for his best-selling account of the voyage, and overseeing paintings by Samuel Scott of Anson's naval career. He was appointed captain of the *Royal Caroline* in 1752 and was knighted after the King's voyage in 1752. He was clearly not expected to devote his time to his yachts: he took part in a survey of the port of Harwich in 1754. Nor was he expected to serve out his time in them. He returned to active service in 1758, serving as first captain, that is chief of staff, to Anson in the *Royal George*.

William Browell was the son of one of Anson's midshipmen in the circumnavigation, which guaranteed a good start to a naval career. As a lieutenant he fought at the Battle of Dogger Bank in 1781, took part in the occupation of Toulon in 1794 and the Glorious First of June battle in the following year. He was promoted captain to take command of the *Princess Augusta* but moved on soon afterwards, becoming flag captain to Lord Hugh Seymour.

By the end of the century the captains were even more likely to be officers of some distinction. Commodore Sir Charles Paget was the brother of the Marquess of Anglesey. After a distinguished naval career from his first entry in 1790, Charles had a strong connection with George as a Groom of the Bedchamber, and had been knighted by him in his favourite resort of Brighton in 1819. He was appointed to the *Prince Regent* in 1819 and the *Royal George* in the following year. He was promoted to the temporary rank of commodore in July 1822, just before the voyage to the north.

Charles Adam was well connected, the nephew of Admiral Lord Keith, one of the wealthiest men in the navy. Charles entered the navy in 1790 at the age of ten and during his teenage years his uncle (who was generally reckoned a good judge of character) had him promoted to acting captain by sixteen. He was brought to earth when he was summoned to the Admiralty Board, which pointed out that his under-age promotion was illegal and he was reduced to midshipman. He quickly rose through the ranks again and co-operated with the Spanish guerrillas off the coast of Spain in 1810. He took command of the *Royal Sovereign* in 1821 and according to Lieutenant Boteler he was 'more like a midshipman than a staid post captain. He entertained and amused us greatly'.⁸

Captain the Honourable Robert Cavendish Spencer was the second son of Earl Spencer, who had served as First Lord of the Admiralty in the 1790s, and was responsible for sending Nelson to his first great triumph at the Battle of the Nile in 1798. Robert had entered the navy in 1804 at the usual age of thirteen and was on board the *Tigre* during Nelson's epic chase of the French fleet across the Atlantic, though he did not serve at Trafalgar. He was in command of the 16-gun *Kite* by 1813 and as captain of the *Carron* he took part in operations against New Orleans. He commanded the *Royal Sovereign* during Clarence's escapade of 1828.

Captain the Honourable Robert Cavendish Spencer, by Maxim Gauci and Thomas Phillips. (© National Maritime Museum, Greenwich, London, PAF3541)

Warrant officers

Under the captain, each yacht had a hierarchy almost exactly like a ship of the line or frigate, which meant a high proportion of warrant officers. They were distinguished from commissioned officers in that they were appointed by the Navy Board rather than the Admiralty, and they had specialised functions. They came in three classes. The higher ones, masters, surgeons and pursers, were almost equal to commissioned officers in status. The middling ones, boatswains, carpenters and gunners, had invariably risen from the lower deck, but had much status and power aboard ship. The lower grades included clerks and cooks who were little distinguished from the crew, except that they held their ranks permanently. However, the cook appointed to the *Royal Sovereign* in 1822 was different, according to Boteler. 'I took him for a gentleman, said he was a cook, so I sent him down to his dominions. … He was a first-rate artiste and it was a sight (with his white dress, cap also, and with a belt and knives in it) to see him handle pastry etc.'[9]

The master was a very experienced seaman, the deputy to the captain in running the ship if no lieutenant was appointed. 'But as the master is the principal officer next under the captain, has the charge of piloting the yacht, must keep a journal and log book and take account of the provisions and be placed in a readiness to be upon deck at all calls'.[10] In 1749 Captain Thomas Limeburner of the *Fubbs* claimed that the master was qualified to take the yacht to Holland if necessary.[11] Franklyn, master of the *Royal Sovereign* in the 1820s, was a 'fiery-eyed Welshman' who could be mistaken for a mental patient.[12] In 1728 the Admiralty decreed that the *Carolina*, *William and Mary*, *Fubbs*, *Mary* and *Katherine* should each have a second master instead of a master's mate, tending to spread the responsibility.[13]

The appointments of surgeons to the yachts were rather desultory in the early days. The *Mary* had none until 1666, then three in succession by 1670. The *Anne* had none until Thomas Stephenson was appointed in 1670. Henry Walker was appointed to the *Katherine* in 1666 to be succeeded by John Harrison in 1670. At the time surgeons were inferior to the physician and were trained by apprenticeship rather than at university. They were expected to find their 'necessaries' for the treatment of patients, but the government was often late in paying them. In 1683 Ralph Leage of the *Cleveland* complained that he had not been paid for seven years.[14] Thomas Tongue, surgeon of the *Peregrine* in the early 1700s, had a long-standing dispute with William Joy, the carpenter of the *Mary*, who accused him of mistreating a wound some years earlier, but he remained in place to bring the new King from Hanover in 1714.[15] In 1721 Richard Beale applied to be surgeon of the *Katherine*, listing six ships he had served in, in the Mediterranean and Baltic and at Newfoundland.[16] By the time he died in 1731 he was surgeon of the *Caroline*.[17]

In the early days the yachts were administered by clerks, leaving a great deal of responsibility to the captains. In 1726 Isaac Stacey of the *Charlotte* was suspected of forming part of a ring which received money from the Pye Alehouse. He made a statement implicating the master, carpenter, boatswain and a midshipman in the affair.[18] In later years pursers were appointed as supply officers. Thomas Goddard entered the navy in 1787 to work in the office of Admiral Sir William Parker and he served three other admirals over the next seven years. As purser of the *Apollo* frigate he was wrecked off the Scheldt in 1794. There were two more appointments as admiral's secretaries, and off Spain in 1810 he purchased provisions for the support of Lord Exmouth's squadron. As a diversion, he 'compiled and delivered to the … the First Lord of the Admiralty a classed register or pedigree of each ship then in the navy, setting forth her sailing and all other qualities'. In 1812 he was appointed to the 98-gun *Trafalgar* building at Chatham, something of a sinecure, and then for the first time to the *Royal George* yacht. He would continue to be reappointed until 1831, apart from two years in another *Apollo* frigate, but for most of that time he was absent, setting up chains of signal stations between the Admiralty and the main naval bases. In 1834, at the age of sixty-two, he was reported to be 'Able and willing' to serve at sea if called on and he was still purser of the *Royal George*.

The three warrant officers on the second level were the specialists, the boatswain, gunner and carpenter. They were known as the standing officers, because in a regular warship they stayed with the ship when she was out of commission – which did not normally apply to the yachts. They tended to have close links with their families in Deptford. In 1682 John Evaines, the carpenter of the *Navy*, was concerned that the yacht might be transferred to Portsmouth and his would lose contact. In 1714 Nicholas Voden, the carpenter of the *William and Mary*, asked for permission to sleep at home to look after his children – though it seems likely that others did so without asking. In 1802 James Mackie, gunner of the *Royal Charlotte*, was reported to have built a lean-to on the wall of the dockyard, suggesting he had not lost contact with the shore.[19] The boatswain was responsible for the rigging, boats and anchors of the ship, as well as mustering the crew, allocating their duties and maintaining discipline. The carpenter maintained the hull and masts; the gunner was responsible for the guns themselves, as well as their carriages, tackle and powder, which was perhaps the heaviest of his responsibilities.

Lieutenants and midshipmen

The captain of the early yachts only had the master who was fully qualified to share watch-keeping duties, which could work well on a short voyage but caused difficulties on a a longer one.

In 1682 Captain Gunman left the master's mate William Sturgeon in charge during the night in which the *Gloucester* went aground. He was not censured for this by the court martial, so perhaps it was considered normal. By the 1770s it became normal to appoint a lieutenant as second in command of a royal yacht, relieving some of the pressure on the captain and master on a prolonged voyage, such as Sandwich undertook during his visitations. George Robinson Walters was appointed to the role in the *Augusta* in 1773, but only on a temporary basis; Jemmet Mainwaring held the post in the same ship, now renamed, in 1795. When John Harvey Boteler was appointed to the *Royal Sovereign* in 1822 he found it a 'most gratifying and pleasing appointment'. When he arrived at Deptford Dockyard gate with three midshipmen, 'no small stir was made when we announced ourselves officers of the yacht, the warden himself, an ancient lieutenant, offering to shew us the way.'[20]

Midshipmen had originally been petty officers, but during the eighteenth century they gradually became established as trainee officers, and as such they were given a blue uniform in 1748 along with other officers, with white patches to distinguish them. There was usually one to be found on board each yacht in the eighteenth century – Joseph Amey in the *Fubbs* in 1722, Thomas Smith in the same ship in 1765 (though at the age of forty-four he was probably not to be taken seriously as a trainee officer). By the end of the century there was a tendency to increase the numbers. In 1773 the *Augusta* had two, including Gabriel Bray, who would later produce fine watercolours of shipboard life in a frigate. All three of the midshipmen of the *Princess Augusta* in 1795 had good naval connections, being the sons of captains Hammond, Christian and Maud. By 1721 the *Royal George* had four 'Admiralty midshipmen' who were presumably appointed by that board rather than the captain of the ship. However, they included Lord William Paget, who was the son of the captain, as well as Courtenay Boyle, the son of an admiral and grandson of the Earl of Cork and Orrery. The complement would normally include one or more captains' servants, whose status is often difficult to determine. The term might be used in the sense of an apprentice, and describe a young man serving time needed to become a midshipman, or it might describe a domestic servant. In the *Augusta* in 1773, however, Henry Wray had the same surname as his captain, which is something of an indication.

~ Crews ~

Below the warrant officers was another hierarchy of petty officers, usually promoted from among the seamen. In the *Fubbs* in 1722 George House was the quartermaster, supervising the steering and helping with the navigation under the master. James Brown was promoted from able seaman to boatswain's mate in February. His duties were to assist the boatswain with the rigging and ship's discipline, including flogging offenders, though there is no sign that there was much of that in the yachts. Joseph Thompson was the carpenter's mate and presumably a skilled shipwright. John Spiller was the gunner's mate. Unlike warrant officers they had no security in

Gabriel Bray, who served in the *Princess Augusta* before being appointed to the frigate *Pallas,* where he painted this self-portrait. (© National Maritime Museum, Greenwich, London, PAJ2024)

their rate. In 1772 51-year-old Thomas Kindred of the *Augusta* was reduced from gunner's mate to able seaman and replaced by Thomas Willson. The petty officers of the *Princess Augusta* in 1795 were Edderly the master-at-arms (a rate which was not common in the yachts, and in charge of the ship's discipline), Price the sailmaker, Carpenter's Mate Watts, Boatswain's Mate Nash and Gunner's Mate Benjamin Blackford.[21]

Nearly all the seamen listed in the muster and pay books of the yachts were rated 'able', that is they had a number of years of experience and could 'hand, reef and steer'. In 1713 the *Peregrine Galley* was exceptional in having four 'ordinary seamen' who only had limited sea experience, but at the time she was not yet registered as a royal yacht.[22] There is no sign of any unskilled 'landsmen' in the yachts' crews, though later, boys were taken on – they were capable of learning to become seamen in due course, and in the meantime they were useful as servants to the officers. The average age of the yachts' crews was probably higher than the run-of-the mill seamen. In 1795 eight of the *Princess Augusta's* petty officers and seamen were aged fifty or over, and three more were over forty.[23] Like the warrant officers above them, they had strong links with the shore. In 1773 twenty of the crew of the *Augusta* gave their 'place and county where born' – not just their residence – as neighbouring Deptford, Greenwich, Rotherhithe or Woolwich.

In the early days at least, the crews of the royal yachts were not in any sense an elite, and had their share of brushes with the law. It was reported from Woolwich in 1664: 'Took the servant of the *Henrietta's* carpenter, stealing old iron. Put him in the stocks and threatened him with whipping, when he confessed that he sold it to one of the smiths in the town. Requests a warrant for searching his house, and those of the rest of that trade.' In 1703 William Sanderson advertised a guinea reward for a trumpeter who had deserted from the *Peregrine Galley*: 'John Corell, a fat man, fair haired, of a ruddy complexion, about 30 years old, a trumpeter'.[24]

When Boteler joined the *Royal Sovereign* in 1822 her crew consisted of 'dockyard riggers, old men-of-war's men, thorough seamen, of good character, and mostly married men.' They proved to be 'most able and willing hands' while tacking up the Scheldt, but that did not stop most of them getting 'stupid as owls, from the lot of Holland's drunk' after visiting Antwerp. And it did not allow for the presence among them of a scoundrel who ran away with money collected for a leaving present.[25]

The later yachts

By the late eighteenth century it was no longer common to keep the yachts fully manned even if there was no work for them. When Sandwich sailed to Portsmouth in the *Augusta* in 1773, ten men were sent on board from the *Royal Charlotte*. In 1795 the manning situation was more critical in wartime and ten pressed men were sent from the depot ship *Sandwich*, with not entirely happy results. By 1821 the crew of the *Royal George* was listed in several parts. First, there was a group of nine headed by the captain's coxswain and including four able seamen, plus cooks and stewards. Presumably they were intended to be on board at all times. Secondly, there was the captain and one or two lieutenants, who probably had other duties when not needed for the yacht. 'List no 3' included five warrant officers – the master, purser, surgeon, boatswain, carpenter, but not the gunner, who was listed elsewhere. It is clear that the warrant officers were not expected to be in constant attendance, and indeed Joseph Goddard the purser was fully occupied in setting up the signal stations. There seems to have been a strong element of training: the pay book includes four midshipman and six 'boys first class', one of whom deserted. When needed, the rest of the crew was made up of 'supernumeraries for wages and victuals' who were lent from other ships.[26]

The household and servants

Members of the royal household and servants were not normally listed in the yachts' muster books in the earlier decades of the eighteenth century, presumably because they were only on board for a short time and had their own supply arrangements. However, in 1773 Sandwich's entourage included five servants who boarded at Deptford. During the *Princess Augusta*'s sojourn of 1794/5 the servants were on board for a long time. We know that Richard Miles (1) and Thomas Goodall dined with the petty officers. Above them on the list were David Rice the Second Under clerk of the Kitchen, William Cope the First Groom of the Kitchen and William Donaldson the Second Yeoman of the Kitchen, who perhaps dined with the officers. The status of Thomas Aldon is not clear. Below Rice and Donaldson were Richard Miles (2), John Smeaton, William Smith and G E Hammond, who were perhaps too humble to be named on the lists; they probably dined with the crew.[27]

~2~
Design and Build

~ Designers ~

The design and the building of ships were inseparable in the seventeenth century – the same master shipwright would normally do both. The most prolific Stuart yacht designers were the members of the Pett family, descendants of Phineas who had designed and built Charles I's great ship, the *Sovereign of the Seas*, in 1637. Christopher Pett (1620–1668) was the eleventh son of old Phineas. He played a leading role in designing the new frigates which became the principal ships of the navy in the 1650s. After that he designed the early yachts *Anne* of 1661, *Charles* of 1662 and *Henrietta* of 1663. Peter Pett I designed the *Katherine* of 1661 and the *Jemmy* of 1662. Phineas Pett II built the new *Katherine* of 1672, the *Mary* of 1677, the *Fubbs* of 1682 and the *Isabella* of 1683. Phineas Pett III designed the *Portsmouth* of 1674 and the *Charlotte* of 1677. No plans of their ships survive except in the form of later rebuilds, and they left no treatises on naval architecture – indeed, they tended to be secretive about their craft. As a result, it is difficult to know much about their methods, though they seem to have satisfied the King, who went back to order more yachts from them.

Samuel Pepys, a rising star of the naval administration in the 1660s and 1670s, resented the power of the Pett family and promoted his friend Anthony Deane to counter it. Deane designed the *Cleveland* of 1671, the *Navy* of 1673 and the *Charles* of 1675; his *Greyhound*, often used as a kind of substitute royal yacht, was highly regarded by all concerned and established his reputation for fast ships. According to Pepys, Deane was 'the first that that come up with any certainty beforehand of foretelling the draught of water of a ship before she is launched', but his originality in that respect is doubtful. But according to the shipwright and author William Sutherland, 'I could never learn that Sir Anthony was much of a mathematician, or a very great proficient in the practice, but had the art of talking well'. Sutherland favoured Phineas Pett: 'Sir Phineas was the greatest proficient, since he produced what he did from rules, which he would have made general had he survived with others, but the end of King James was the end of all the said gentlemen, leaving the shipwright's art as far from exact standards as to shape and dimensions as it was in its infant state.'[1] Rivalry between shipwrights would remain a constant feature of this century and the next.

The building of yachts was not a naval monopoly in the seventeenth century. William Castle of Rotherhithe built the *Kitchen* in 1670. Thomas Shish of Woolwich built the second *Henrietta* in 1679 and Phineas Pett II, though still associated with the royal dockyards, built the *Fubbs* and *Isabella* privately at Greenwich. The Marquess of Carmarthen, of course, had strong views on ship design, but apart from the well-known keel, the *Peregrine Galley* was quite conventional in her shape.

After that there were no entirely new major yachts for half a century but Sir Jacob Ackworth, the domineering Surveyor of the Navy, may have played a part in rebuilds. Richard Stacey, the cantankerous master shipwright at Deptford, rebuilt the *Katherine* in 1720, the *Fubbs* in 1724 and the *Royal Caroline* 1733. Ackworth's co-surveyor from 1747 and later his successor from 1749, Joseph Allin, organised a competition for a new yacht but, unsurprisingly, won it himself. His designs for warships are generally considered uninspired, but the *Royal Caroline* of 1749 was one of the most successful of all the yachts.

With the revival of yacht building early in the nineteenth century, it was common for the Surveyor of the Navy rather than the master shipwright of the yard concerned to design ships, while the master shipwright of the dockyard supervised the construction. As a result, Sir John Henslow designed the *Royal Sovereign* of 1804 and the *William and Mary* of 1807, and Sir Henry Peake the *Royal George* of 1817. The *Prince Regent* was designed as an exercise by the students of the School of Naval Architecture at Portsmouth in 1820. That institution was abolished when Sir William Symonds, a naval captain rather than a shipwright, took over as Surveyor in 1832, to the outrage of the shipwright profession. His talents might have been useful in designing a fast seagoing yacht, but his only contribution to the genre was the *Royal Adelaide* miniature frigate for Virginia Water in 1833.

~ Drawing the profile ~

Anthony Deane provided the most coherent English treatise on ship design of the Stuart period, with his *Doctrine of Naval Architecture* of 1670. He had not produced any yachts by that time and instead he concentrates on a third rate ship of 70 guns, but there is no reason to believe that the principles were any different. The basic dimensions – length on the keel, maximum breadth and depth in the hold - were usually given when the vessel was ordered, though in the seventeenth century it was quite common for the shipbuilder to increase them slightly during building in order to claim extra payment for tonnage.

Deane, like every other shipwright of the age, started by drawing the keel, which was invariably a straight line, and its length was one of the principal dimensions of the ship. He wrote, 'The first stroke that is struck in this work is the line of the keel'.[2] At its after end the stern post rose at a slight angle from the vertical, around 12 degrees, though the new *Royal Caroline* of 1749 had a slightly more vertical post, perhaps following the fashion set by captured French ships of the time. Of the nineteenth-century yachts, the *William and Mary*, *Royal George* and *Royal Charlotte* also had posts about 10 degrees from the vertical, while the *Royal Sovereign* and *Prince Regent* had the more traditional 12 degrees.

The bow was a curve when seen in profile. The most common method of drawing it was a circle tangential to the keel and ending almost vertically about the level of the main deck. The *Henrietta* of 1679 had a more upright stern made up of a small and a large curve, which produced more internal space. In the *Fubbs* as rebuilt in 1724, and possibly in the original, it met the keel at an angle, which produced a shorter bow. Conversely, the *Peregrine Galley* bow was tangential to the keel, but the circle had a much larger radius so that it did not reach the vertical on the level of the main deck. This was described as a 'scow bow' by a modern naval architect and it had much influence, though it did not displace the more conventional bow. In any case, forward of the bow was the knee of the head, in line with the keel, which supported the figurehead and its rails. The designer then added the decks at suitable heights, with partitions for the cabins. In the yachts, of course, the main task was to provide a large stateroom at the stern for the King or distinguished passengers.

The midship section

After the sheer plan, the shipwright drew the midship section, which usually had a straight line, known as the floor, coming out from the keel, with a series of tangent arcs above to form the shape. Presumably the first *Mary* had a typical Dutch flat bottom, which is why she needed leeboards, and why she was chosen to operate in the shallow waters of the Dee Estuary. However, the evidence is strangely contradictory, Pepys lists her draught as 10ft, 2ft more than any other yacht, which might be explained if he was including the leeboards in their down position. Anthony Deane gives her 7ft, equal to the *Anne* and not much more than the other yachts, which is even harder to explain.[3] In any case, the English yachts soon moved to a very different midship section, according to the evidence of models. The early ones have a flattened, 'V'-shaped lower part, known technically as rising floors. This would produce a deep hull for a given size and would prevent it being blown sideways in the wind, but it restricted space for ballast and stores in the hold. Perhaps this is why it was so important to find lead ballast for them.

Above the floors, whether rising or not, there was a relatively small curve known as the floor sweep and a roughly similar curve at the broadest part of the ship just above the waterline. These were joined by a larger curve known as the reconciling sweep. Above maximum breadth the hull narrowed in a process known as tumblehome. Usually this was done by a conventional curve, the above breadth sweep, and another curve in the reverse direction, the toptimber sweep. This produced the classic 'tulip' midship section, characteristic of British warships of the age, though it was slightly flattened in the case of the yachts as compared with two- and three-deckers. The *Katherine* as rebuilt in 1720 had a wider hull at the level of the waterline, producing a fatter appearance when seen from astern.[4]

Tumblehome tended to be reduced over the decades in all ships, producing a wider deck which might benefit the passengers of the yachts. It was quite small in the *Royal Sovereign* of 1804, which also had quite steep floors. This was developed further in the *William and Mary* of 1806, which had an almost 'U'-shaped section but with a slight 'V' bottom. The 'V' was more pronounced in the *Royal George* of 1817, and the shape was almost duplicated in the *Prince Regent* of 1820, and the *Royal Charlotte* of 1824, though they were all designed by different naval architects.

The lines of the hull

Now the shipwright had to form the shape of the hull before and aft of the midship bend, a mystical process to some. Deane conventionally used curved construction lines, the rising and narrowing lines of the floor and maximum breadth, to achieve this. Aft, the hull had to be narrowed sharply to allow the water to reach the rudder. This could be taken too far in a warship, which had to have enough buoyancy in that area to support the weight of the guns. It was slightly easier in yachts, which did not have guns aft, but it might produce too narrow a space under the stern, which restricted the accommodation.

When building the *Fubbs* in 1682, Sir Phineas Pett used the highly technical method of the Cono-cuneus, or the solid of least resistance. He had collaborated with the mathematician John Wallis on developing this, and was acknowledged in his book on the subject, dedicated to Sir Robert Moray: 'Since I came home from London, I have taken for some time to consider those solids and lines made by the sections thereof, proposed to your consideration … by Mr Pett, one of the Commissioners of His Majesty's Navy.' But it was beyond the ken of the great majority of shipwrights, who had only an elementary knowledge of mathematics; and even William Sutherland had to admit that it had little use in practice. The bows were much bluffer than the stern, with the shape partly dictated by the curve of the stem post. Apart from that, the main difference over the years was the amount of hollow in them.

Having drawn out the shape of the hull using tangent arcs, and curves or straight lines to join them to the keel away from midships, the shipwright attempted to 'fair' it out and remove any local bumps and irregularities by drawing horizontal lines (waterlines), vertical (buttock) lines and diagonals.

The building slips and dry dock attached to the wet dock in Deptford Dockyard, where yachts were often built or repaired. (© National Maritime Museum, Greenwich, London, SLR2906)

⁓ Construction ⁓

There is little detailed information on the construction of the early yachts, but it is likely that it was conventional, and there is no sign that any of them were experimental in that respect. Shipbuilding contracts gave full details of the structure, but there are none for the yachts, which were mostly built in the royal dockyards. The Establishments of Dimensions, especially that of 1719, gave full dimensions for every part of regular warships, but did not reach down to yachts, so we can only glean a few details from letters and surviving plans.

Yachts were usually built on a slip sloping gently towards the water, for dry docks were precious and reserved for repairs, or building the largest ships, such as HMS *Victory* in 1765. At Deptford, there were two slips facing into the great wet dock. The basis of the construction, the keel, was straight and approximately square in cross section, except for the *Peregrine Galley*. It was made in several sections scarfed together – the *Mary* had at least three. The stem post on which the rudder was hung was straight, the bow was formed by a curved stem post which might take various forms as described above. The angle between the stem and stern posts and the keel was filled with flat timber known as the rising deadwood.

The frames were attached above and outward of the keel. It was impossible to find a single piece of timber whose grain would match the curve of a frame, even in such small ships, so they were made in several sections. The floor timber of each pair of frames stretched across the keel and ended at about two-thirds of the maximum breadth. The whole structure was locked together by the kelson above the keel and rising deadwood and that was fitted with sturdy 'steps' to hold the heels of the fore and mainmasts. The first foot-hook or futtock overlapped with that and formed the curve of the lower part of the hull, or the turn of the bilge. Large ships had up to four futtocks overlapping one another, but the *Royal Caroline* of 1749, one of the biggest of the yachts, only had two. The midship frame was at the widest part of the ship and the frames gradually got smaller fore and aft of that. Close to the bow and stern there were half-timbers, attached to either side of the rising deadwood. From 1715 onwards they tended to angle increasingly forward towards the extremities of the ship. The foremost part was formed of hawse pieces, angled forward and parallel to the line of the keel when seen from above.

The Deptford officers went to unusual trouble to find the best timber for the frame of the new *Royal Caroline* in 1749.

> … most of the our piles of timber, converted and rough, have been searched and have supplied all the rising timbers except O and L, and floor timbers except 8, 10, 11, 13, 14, which we are rummaging for; but find 18 lower futtocks of the roundest sort of the after body and 12 of the fore body not to be had here, made difficult by their figures and length reaching down to the deadwood … and the second futtock coming up to the port sill, the ship having but two futtocks hardly find any to answer the mounds in this yard, have prevailed on the master attendant to let the Lyon hoy, who has sails on her for Chatham, to sail this afternoon on board which I shall put moulds for these difficult timbers …[5]

The frames on each side were linked by deck beams, which only had to carry light guns at most and were not as strong as those of warships. They were braced by knees, mostly shaped like inverted 'L's. The decks were planked, apparently 2in thick on the lower deck of the old *Royal Caroline*, and 3in on the upper deck. The exterior of the bottom had to be kept smooth and it was covered in 3in plank. The three strakes just below the widest part of the frame widened gradually to 5½in, and above that were thicker pieces known as the wales. These were double in the early yachts and in the old *Royal Caroline* the lower one was 6in thick and the upper one was 5½in with a 3in strake in between. This feature was retained in the new *Royal Caroline* and even on the *Augusta* of 1771, though it was very old-fashioned by that time; later yachts had a 'single' wale, with all parts of the same thickness. The planking above that varied from 3½in to 2in and it was topped by a flat, horizontal gunwale. Internally the planking was thicker over the joins in the futtocks, and under the decks.

The external planking was 'dubbed' with an adze to provide a smoother surface. It was caulked by driving oakum

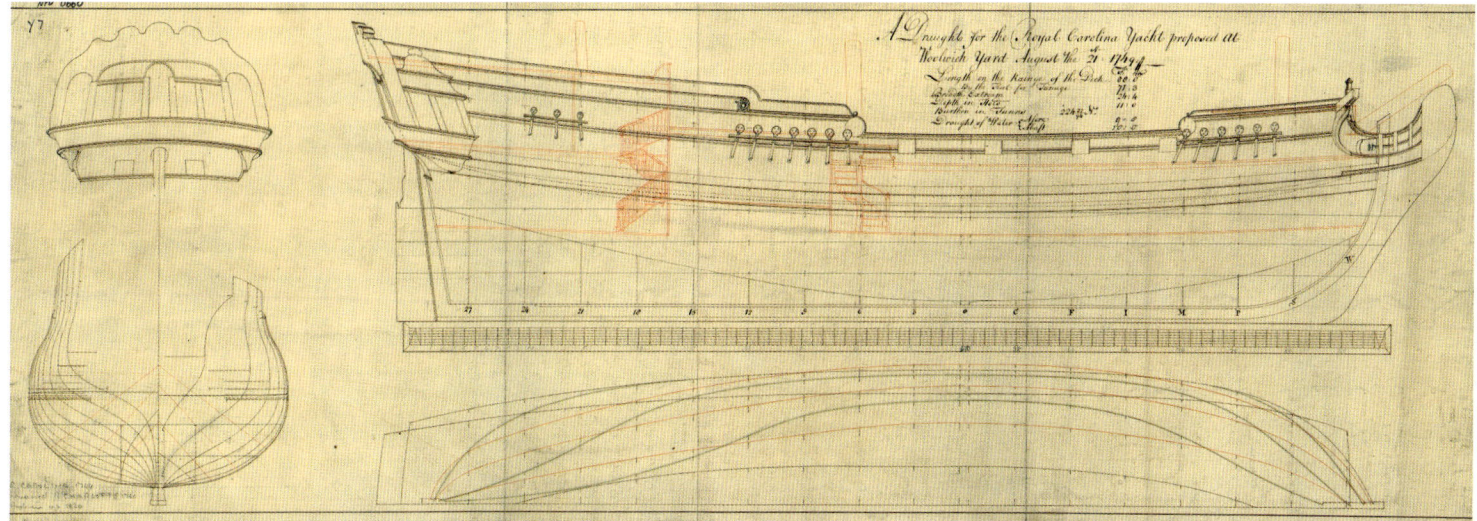

into the gaps between the planks and then covering that with tar to seal it. A yacht might well be tallowed to make the surface smoother. When ready, the yacht had cradles set under her bow and stern, set on rails to allow her to slide into the water. The launch was rarely the subject of much ceremony at that time, though Charles did attend that of the unsuccessful *Isabella* in 1682. In 1804 the *Royal Sovereign* was launched 'amid the plaudits of some thousands of spectators', and William IV had some ceremony for the launch of the *Royal Adelaide* into Virginia Water. After launch a yacht, like any other vessel, was hauled alongside the sheer hulk in the river to have her lower masts fitted and the rest of her preparation was done by her own crew, perhaps with the aid of dockyard riggers. After that she was ready for duty.

Right: The midship section of the *Royal Caroline* showing details of construction. (© National Maritime Museum, Greenwich, London, NPD0659)

Below: The framing plan of the *Royal Sovereign* as sent to Deptford in November 1801. (© National Maritime Museum, Greenwich, London, NPD0124)

The draft of the *Royal Caroline* as drawn in August 1749. To the left, a drawing of the stern, with the body sections below – aft of midships to the left, forward to the right. The man drawing is the sheer plan or side view, with a certain amount of internal detail including stairways. Below, the waterlines at various levels showing the 'fairness' of the hull which give some indication of sailing qualities. (© National Maritime Museum, Greenwich, London, NPD0660)

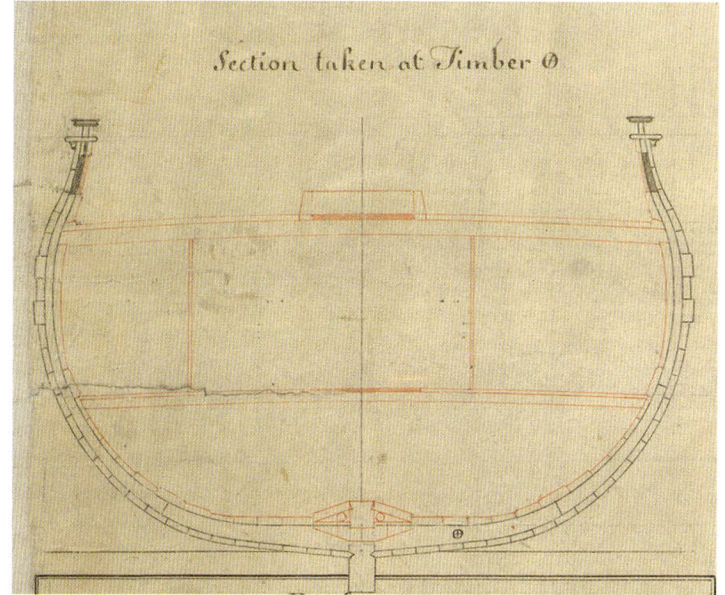

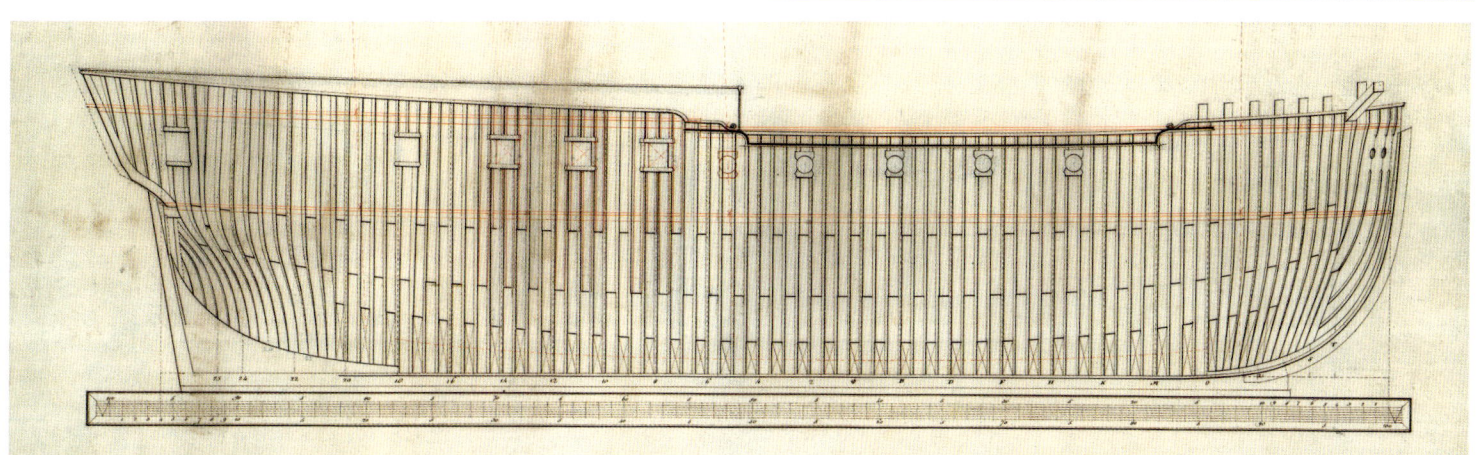

DESIGN AND BUILD

3
Accommodation

Though they developed over the years, the yachts shared certain common features which were rarely found on other vessels. From the seventeenth century onwards, regular warships usually had continuous decks, curving upwards towards bow and stern, but essentially on the same level. This was not possible on the yachts, and decks had to be arranged at different heights to take advantage of the spaces available, provide comfortable accommodation for the passengers, and avoid the very narrow parts of the hull below. The Coronelli print of the 1690s, probably of the *Isabella* of 1683, gives us our first glimpse of the interior of a royal yacht. Towards the stern she had upper cabins on two slightly different levels, with a more spartan one below. Forward of that there are enclosed apartments on only one level, with a flush upper deck and two breaks in the lower deck for the three compartments. In the *Augusta* of 1771 there were several levels for the after cabins, but continuous decks forward of the mainmast. By the early 1800s the middle deck, where most of the royalty and nobility were accommodated, was flush, though the lower deck still had small variations in levels to deploy the space available to best advantage.

There was very little room below decks to store food and drink, and clearly the yachts were not expected to be out of contact with the shore for long. When they were anchored off Sheerness to ride out a storm during William and Mary's passage in 1677, they had to be supplied from the shore. Even the *Royal Caroline* had to go on short allowance during the ten-day passage with the future Queen Caroline in 1761 – in a regular warship that would only happen after months at sea.[1] While the *Princess Augusta* was stuck in the Medway and Nore, she had to go back to Sheerness for more fresh provisions on 1 January 1795, but by the 21st she was 'scarce of several necessities'; on the 25th they put planks over the ice so that the crew could 'get on shore for more provisions'. They did the job well: on the 30th they had 'plenty of the best of drink and provisions.'[2]

All the royal yachts had winding staircases between the most prestigious apartments, with balustrades which must have been necessary in a rough sea. More functional spaces had simple ladders, and the transition might occur in a single flight – in the *Royal Caroline* the curved staircase down from the middle deck turned into a much plainer one for access to the hold. And there might be class distinction when they were side by side: the *Royal George* had a slightly curved stairway for the officers between the lords' and the seamen's compartments, and straight one for the seamen. The *pièce de résistance* of the *Royal Sovereign* was the 'winding staircase whose strings and hand rails, being glued up in thickness and constructed upon a much larger cylinder, with proper framing to receive the treads and risers', which was much more difficult than the one in the old yacht.[3]

Class distinctions

Witsen described the *Katherine*'s interior layout in some detail, highlighting the stark contrast between the royal apartments and the spartan crew's quarters:

> One goes down some steps into two rooms aft, the foremost is a brave hall, painted all round with works of art, gilded and decorated with carvings; the after cabin which one enters from the other is up a step or two; here there is a nice bedstead, and it looks out aft with windows being decorated all round with gold leather. Above this room, which is raised some feet above the deck, stands the helmsman; the iron tiller is conveniently bent so that it can be handled by a man standing upright. The after end of the roof over this stern cabin is higher than the fore end for the sake of appearance.
>
> Below these two rooms in the bottom of the ship are two spaces for stowing rigging gear and tools. In the fore hall is a staircase with a door by which one goes down into a room whose floor is the ship's bottom; through this one enters another, and thus reaches the fore part of the ship; these rooms have beds all round separated by screens painted with the King's arms and gilded.[4]

At the stern of the ship in the Coronelli print, probably the *Isabella* of 1683, there is a richly decorated cabin with a small compartment aft of that, perhaps a toilet. It is dominated by a large four-poster double bed, which fits the image of Charles II, if not William III. Forward of that is an open space with small compartments on either side, probably reserved for the principal guest and his followers. Below that, in the only part of the yacht with two decks is what seems to be the officers' quarters.

Forward of that is a well-fitted room with an oval table in the centre and four paintings on each side, clearly intended for the distinguished passengers, and perhaps the origin of the 'lords' room' in later yachts. There are lockers round the sides, and one is shown angled inward perhaps for access, or to turn it into a bunk. The next space, forward of the mainmast, is much more functional. It is the galley, fitted with shelves, racks and cupboards for storage – for there was practically no hold underneath. The stove itself is not visible, but its chimney can

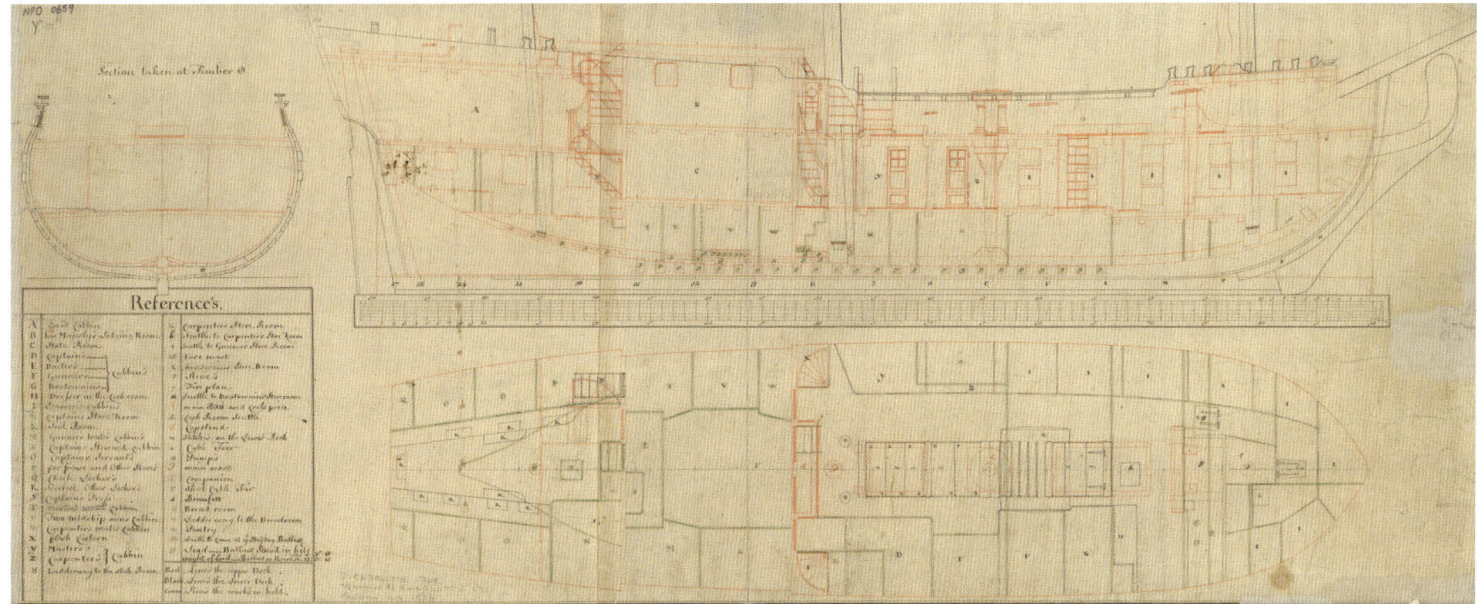

An interior plan or longitudinal section of the *Royal Caroline* of 1749 showing details of the cabin layout and staircases. (© National Maritime Museum, Greenwich, London, NPD0659)

be seen at the forward end of the compartment. Finally, there are the crew's quarters in the bows.

The *Royal Caroline*

Information is scarce for the early eighteenth century, but around 1730 the stern of the *Royal Caroline* was on three levels. At the top was the quarterdeck, open and used to steer the ship, as well as for royal promenading. Below that was the great cabin for the royal accommodation, probably a sitting room. There was another great cabin below, rather restricted by the narrowing of the hull below the waterline. Below that, and more functional, was a small space filled with lead ballast. That took up about 22ft, or a quarter of the length of the ship on the main deck. Forward of that, about 15ft 6in fore and aft, there were also three decks which did not coincide with those aft. The old awning of the Stuart period had developed into a solid structure which was permanently covered. 'His Majesty's room in which he lodges', was halfway between the two great cabins aft. It was George I's favourite space on board, and in 1729 the Deptford officers suggested that part of its fitting should be paid out of the royal rather than the naval budget because of that.[5] Below that was the 'state room' for the courtiers. It had two 'settee beds' against the after bulkhead, and round the sides were six 'locker berths'. Below, at the very lowest level in the ship, was accommodation for servants and petty officers and their equivalents – stewards, midshipmen, and master's, gunner's, carpenter's and boatswain's mates placed round a communal gunroom.

Forward of that, and taking up the remaining 49ft of the ship's length, the two decks were continuous. The upper deck was largely used for working the ship, with the capstan, bitts for securing the anchor cable, and the guns. The lower deck contained cabins for the captain, surgeon, gunner and boatswain on the starboard side with the master and carpenter to port, with hatches to the 4ft deep hold in the centre as well as a 'pound for the small bower' anchor. The foremost area, much affected by the curve of the bows, had thirteen 'private men's cabins' on the port side – actually two- and three-tier enclosed bunks. In the centre was the galley stove, with hatches to the coal hole and the gunner's and boatswain's store rooms, less than 2ft deep.

In the *Royal Caroline* of 1749, the officers' and crew's quarters were spartan. Under the after room were tiny cabins for servants, and store rooms. Even the captain, who might have three cabins and an open gallery to himself in a ship of the line, only had a small compartment on the lower deck forward of the stateroom. In a rare example of egalitarianism, there were single, double and triple cabins in the same area for

Vincenzo Coronelli's drawing of the interior of a royal yacht, showing from left to right; crew's quarters, galley, dining room, officers' quarters (below), open deck (above) and great cabin with a magnificent bed. (From his book *Ships and Vessels*)

Detail of a model showing the interior of the *Royal Sovereign*, with the cabins in the stern. Though the exterior of the model seems very accurate, it is possible that some of the internal detail was added later, including the figures. (Mariners Museum)

the master, doctor, warrant officers and eighteen seamen. However, sea time was infrequent compared with warships, and the captain might well have dined in the royal suite. In harbour he might requisition part of it for his use, or spend most of his time ashore.

The plans of the *Augusta* of 1771 do not name the different compartments, but she has two large cabins aft, one above the other. The upper one was presumably for the King's bedroom, the lower one perhaps for the captain or the royal suite. There was another cabin forward of that, at a level between that of the two aft of it. It was probably used for dining and accommodation of the principal guest and his courtiers or attendants. There is a flat deck forward to that, undivided except for some pillars. It is likely that the midships part included cabins for the officers round the sides, and the fore part was accommodation for the seamen. The galley stove was situated in the middle of that area.

Furnishing

Some idea of the furnishing can be seen from a list of what was decayed in the *Charlotte* in 1727. In the bedchamber, a 'green camblet standing bed with curtain and valence', six window curtains and a counterpane were said to be 'worn and very much soiled.' In the 'stateroom and officers' cabins' were a feather bed, two quilts, eight blankets and four 'blue chiney window curtains and valance', which were also defective. The captain's cabin needed two new bed curtains of similar specification, and three window curtains, the 'ordinary cabins' a blanket and three quilts. As to linen, seven flaxen sheets, two coarse napkins and seven 'Holland pillow bears' could be 'converted for towels, dust and tea cloths'. In addition, a quilt and three blankets were needed for the stateroom, along with two flock beds, two bolsters and two pillows for the ordinary cabins. In china, three more teacups and saucers were wanting, as well as a small copper tea kettle and two stone teapots, plus a gridiron, and a small looking glass.[6]

In 1751 a Mr Amery was employed on the same yacht 'in cleaning, repairing and furnishing the upholsterer's goods amounting to the sum of £41.13.1.' That included '[t]en days work myself and man repairing the several beds and pillows in the state room, … Cleaning a large Turkey carpet' and 'repairing, mending and binding it all round' using 14 yards of 'strong worsted binding', as well as 'making, waxing, emptying the old and filling the new feather beds' using 26 yards of binding including 'picking and dressing the feathers'.[7]

Officers' cabins

In the supposed print of the *Isabella* of 1683 ten officers' compartments were sited below the great cabin and they were spartan. There was a small table forward, and what appears to be cabin doors with possibly enclosed bunks aft of that; each cabin would have to be adapted to fit the narrow hull in that part of the ship. During the next half-century or so, the cabins became larger and more regularly fitted. A plan of the *Fubbs*, probably for the rebuild of 1724, shows a compartment in the after part of the lower deck with a central space and four cabins. The after ones on each side are larger, but are clearly much restricted by the narrowing of the hull below. Forward of that there are cabins shown by dotted lines, which perhaps means that they were removable. In the *Royal Caroline* each had a door with a window in its upper half, and a lower section which could presumably be kept shut to keep out any water on the deck.

In the *William and Mary* in 1752, Captain Parry denied that there had ever been 'any absolute establishment for the officers' cabins in the yachts', though the ranks of their occupants were often marked up in plans, which might suggest the contrary. He outlined a game of musical chairs played over the years.

> At the building of that yacht [the *Mary*] Sir William Sanderson appointed the cabin which the master now has to the gunner and the carpenter had a cabin on the other side of the yacht. There was then no warrant master but the mate had the cabin which the surgeon now has. When Captain Robinson commanded the yacht he disposed of the cabins in another manner but I cannot justly tell how. Captain Brett whom I succeeded had placed them thus. The gunner was in the cabin the master is now in and the carpenter still in that which he has now. The gunner soon after being removed into the *Royal Carolina* I appropriated that cabin for the master who before lay in a very inconvenient dark cabin where he could not so much as see to write. … I judged it but reasonable that he should have the most convenient cabin which was then vacant. The carpenter now has the same cabin his predecessor possessed for 17 or 18 years past.[8]

Crew and servants

The crew was invariably accommodated well forward, as was universal in ships of the time, but we only get occasional glimpses of how their quarters were fitted, mostly on plans. The Coronelli print shows seats with enclosed bunks on the sides. The forward compartment of the 1724 *Fubbs* is presumably for the crew, with cabins on each side, probably shared. In 1729 the Deptford officers suggested that eight small cabins should be erected under the stateroom for the seamen.[9] In the *Royal Caroline* they slept in enclosed bunks, with entrances about 2ft square. In the next century there was a move towards a more conventional form of naval accommodation. In the *William and Mary* of 1807 the crew slept in hammocks, and on each side of their compartment there was a 'Wing with balusters in front for storing hammocks'. In the centre of the space there was a 'Bin for seamen's bags'. There was room for three mess tables. This layout was repeated on subsequent yachts. Water closets became increasingly common, and were even fitted in the crew's quarters in the *Royal George* by 1842.

In the *Royal George* of 1817 most of the lower deck stern cabins were allocated to pages, an ambiguous term which might refer to lower servants or to boys of aristocratic origin in the royal service. In addition, there was a cabin for the first steward, opposite his china and linen storeroom. Forward of that was the 'pages' apartment' with the more functional slop room and sail room taking up part of the space. By 1842 the same space in the ship was taken up with 'Livery domestics sleeping cabins'. The '1st steward's china and linen store room' was still in place but his former cabin was now occupied by the clerk of the kitchen.

A standard pattern

By the 1800s, when new yachts were built and more plans are available, a standard layout had emerged for the yachts, as described in press reports on the *Royal George*:

> Yesterday morning we visited the *Royal George* lying off Deptford; she is a noble vessel, with an exterior distinguished for its symmetry rather than for its embellishments. On stepping on board we noticed the same elegant simplicity; with the exception of the gilt coat of arms at one end, the gilt tiller at the other, and a beautiful orbicular glass compass case in the centre, the quarterdeck of the royal yacht scarcely exceeded, in decoration one of our crack frigates.

There was the usual winding staircase, leading from the port gangway and leading down to the King's dining cabin, which was about 23ft wide and 17ft long. A door led to a passage into the King's state cabin or drawing room, about 20ft wide and 15 or 16ft deep, containing two bookcases with a copy of the recently published James's *Naval History*. The royal bedchamber was 13ft by 7ft 6in; on the other side of the passage was 'a small bed-chamber for the personage on board next to the King.' The ceiling was about 6ft 6in high. It was noted: 'the royal apartments, though richly and tastefully fitted up, did not appear to contain a single superfluous piece of furniture.' The fore part of the ship contained the lords' apartment where the captain and officers and some of the principal attendants dined. It was observed that the seats could all be converted into beds, for beyond the royal compartments the accommodation was typically utilitarian and cramped. The kitchen was in the bows, and 'the cooking process was wholly managed by steam'. Below were the sleeping quarters of the officers and the hammocks of the crew.[10]

Aft on the upper deck on one plan of the *Royal Sovereign* was a 'royal apartment', described as the 'Queen's State Room'.

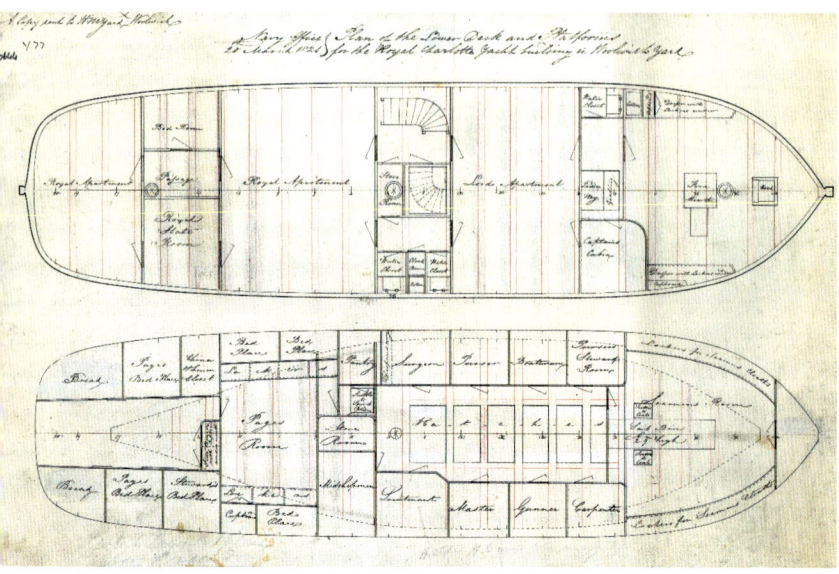

Deck plans of the *Royal Charlotte* of 1824, drawn in 1821. (© National Maritime Museum, Greenwich, London, NPD0844)

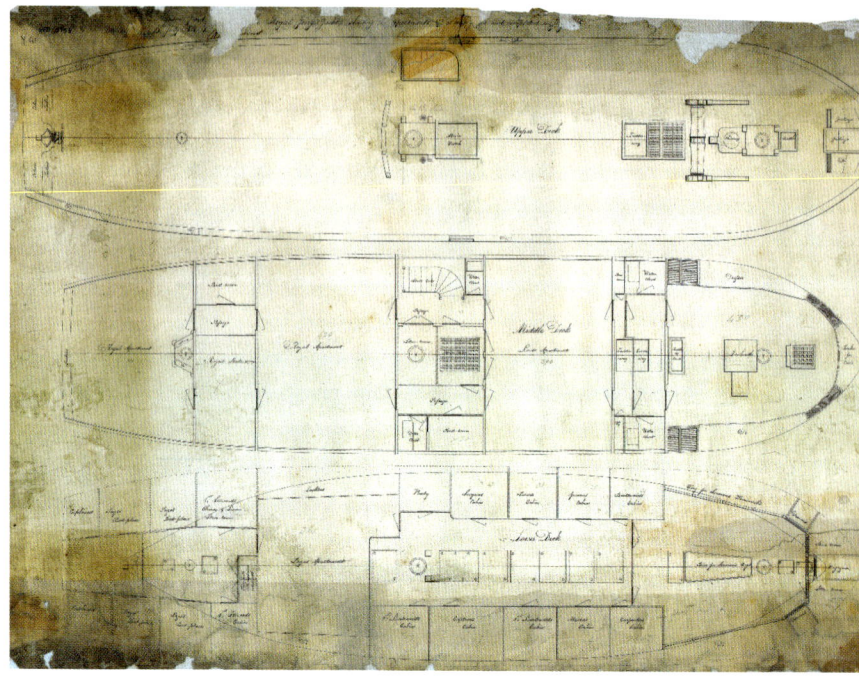

Deck plans of the *Royal George* of 1817. (© National Maritime Museum, Greenwich, London, NPD0733)

It was 16ft by 13ft on the *Royal George*, and of course tapering towards the stern, with an area of 325sq ft. In the *Royal Sovereign* and *Royal George* a space was marked off on each side for a 'small bed place'. The mizzen mast passed close to its forward bulkhead, and it was concealed by two sets of shelves, which were described as the 'Library' by Queen Victoria's time. Forward of that were two bed places of unequal size with a passage between. In the *Royal George* both were 7ft 7in long, while the port one had an area of 32sq ft, the starboard one was 13ft 7in long and contained a water closet. Forward of them was a larger royal apartment or state room, of 418sq ft on the *Royal George*. The next thwartships area was divided into several compartments, including the stairwell and several passages, the main hatch for provisions, the mainmast with a storage area around it, and another water closet. In the *Royal Sovereign*, according to Boteler in 1822, two lady passengers were put up in George III's bed – 'a standing four poster'. They were pitched out in rough weather.[11]

Then came the 'Lords' Apartment', almost rectangular and taking up 385sq ft on the *Royal George*. On either side of this compartment on the *Royal Sovereign* there were '[o]ccasional berths for noblemen in waiting separated with curtains, and a large drop curtain in front to draw up as occasion may require.' The space was 20ft long, which allowed for three berths, each a little over 4ft wide and with a porthole. It was a hardship for those who were used to the comforts of a stately home, but normally it only had to be endured for a few days, and it was the price of royal favour. Evidently there was no room for the paintings which had decorated the equivalent space in the Stuart yachts. When there were no passengers on board, the officers might dine in some comfort, as Boteler noted: 'The officers had just sat down to dinner, and I found the surgeon, purser and clerk at table in the lords' room, where the captain and officers all mess, together with any attendants on royalty that may be on board.'[12]

Forward of the lords' room there were lobbies and water closets, sometimes round a hatch. These provided some distance from the seamen's quarters in the bows. In the *Royal George* and the *Royal Charlotte* the galley was placed in the middle of this space; in the *Royal Sovereign* and *Prince Regent* it was a deck below, with a funnel to carry the smoke through it.

Accommodation afloat is always more cramped than its equivalent on land (except perhaps for the floating mansions of modern billionaires) and the sailing yachts provided an extreme example of this. The seaman was used to swapping his rooms or cottage ashore for a narrow bunk on the yacht, but the contrast was far greater for a king, who exchanged his palace for two or three small cabins, however well furnished. But for most of the passengers the yacht was a vehicle like a coach rather than a residence, and they only expected to be on board for a few days at most – it was different for sea lovers such as Charles II and, more surprisingly, George IV, whose stay might extend into weeks. But none of the sailing yachts was ever to become a home from home, as happened to the *Britannia* in the second half of the twentieth century. The accommodation of the sailing yachts must be seen in that context.

132 ROYAL YACHTS UNDER SAIL

4
Art and Decoration

Artists and craftsmen

The decoration of English royal yachts started off on a very high plane, the pace being set by the standards of the wealthy Dutch institutions and individuals. After her capture, Witsen described the *Katherine* as 'a very costly yacht of His Majesty the King of England', implying that her decoration was lavish even by Dutch standards.[1] English shipwrights of the time were keen to glorify their ships of any kind, and were encouraged by the King, to the horror of the cash-strapped Navy Board. When there was a question about the lavish decoration of the *Mary* yacht in 1677, Samuel Pepys commented, 'the master-shipwrights, when questioned touching their warrant upon which they ground their proceedings on any of these works of charge, do generally pretend his Majesty's verbal command'. The King admitted to the fault:

> His Majesty was pleased … to direct that the same be made known that a standing rule for the future to all the master-shipwrights and other officers in his yards, that he would not expect their putting in execution any verbal orders given them from himself wherein any of his treasure or stores is to be expended or employed, until they shall have communicated such his Majesty's orders to the Officers of the Navy.[2]

Expense on decoration remained lavish in the next century. In 1733 the *Royal Caroline* was repaired at an expense of £7413, more than half of which was on decoration – £1358 for carved work, £1507 for gilding, £38 for painting, £63 for a 'history landskip', and £793 for joiners' work.[3]

Yachts were usually built and fitted at Deptford, which in cultural terms was a long way from the London salons and art galleries and their aristocratic patrons, but it is no coincidence that two or possibly three major artists – Grinling Gibbons, John Cleveley and Charles Brooking – originated there, along with others whose work has been lost; and the Van de Veldes certainly contributed to the yachts in several ways. As the home of the royal yachts, Deptford had a standard of craftsmanship that was unequalled in the other dockyards, and often moved over the boundary into fine art. Figureheads and stern carvings of warships were robust and sometimes crude, but the decorations of the yachts, especially internally, were much more delicate and allowed the finest craftsmanship. Grinling Gibbons is believed to have moved to Deptford in the 1660s from his native Holland and it is not unlikely that he was attracted by work on the yachts. Only the *Katherine* of 1661 was built in the dockyard during that period, but the *Merlin*, *Monmouth* and *Kitchen* were built nearby in Rotherhithe between 1666 and 1670. Gibbons was apparently on other work in January 1671, in 'a poor solitary thatched house' near John Evelyn's mansion at Sayes Court. Evelyn looked through the window and saw him working on a copy of a crucifixion by Tintoretto, 'such a work as for the curiosity of handling, drawing and studious exactness I had never seen before in all my travels'. Gibbons explained that he was working in such an obscure spot 'in order that he might apply himself to his profession without interruption', but perhaps that was not the whole story.[4] In any case, Evelyn introduced him to the King and started off a career as perhaps the greatest wood carver of all time.

Others remained obscure. Henry Turner first appears in the records in September 1714, when he gilded and painted the Admiralty barge and a boat belonging to the *Fubbs*.[5] In 1716 he was paid for work on the new St Paul's Church in Deptford, which may have lured him into the area.[6] In 1718 the dockyard officers had 'no precedent' for work of the quality he did for the King's arms on the yard gate and the *Royal Caroline* yacht, and 'at the mould loft to paint blocks [ie models] etc'.[7] Thomas Burroughs first appears in December 1748, having carved the decorations for the *Lancaster* at Woolwich and the *Culloden* at Deptford.[8] In 1750 he carried out the work on the *Royal Caroline* and justified the cost on the grounds that 'the prices he has charged for each particular are the very lowest he can take for the same and that he was obliged to pay very high wages and to be at great expense in procuring the best workmen from Portsmouth and Chatham to perform the said works.'

Bow and figurehead

The knee of the head projected diagonally forward from the keel of a ship or yacht; it supported the figurehead at its forward extremity, and the curved rails which joined it to the main structure of the hull. It is quite difficult to get information about the figureheads of the Stuart period. Artists, including the Van de Veldes, preferred to draw ships from the stern. When they did portray the bows, the figurehead is often quite indistinct. Models from the period are often difficult to identify with real ships and, moreover, we can never be sure that the designs were executed as portrayed. And in the next century, when more plans become available, the figurehead was usually shown as an amorphous block.

The prototype, the *Mary*, had a typically steep, narrow, Dutch-style head with a unicorn figurehead. According to Witsen, the first English yacht, the *Katherine*, had 'a fine yellow-painted head supported by sea-nymphs and goddesses'.[9] It is not clear exactly what the head represented, but it may have been a mounted figure. The lion was the standard figurehead for late Stuart warships, with a curved and rather emaciated torso and a crown on its head, and indeed such a carving can be discerned on the minor yachts such as the *Merlin* and *Kitchen*. The Van de Velde drawing of the *Charlotte* of 1677 shows an indistinct shape, which may be a woman with angels holding a crown over her head. The *Fubbs* of 1682 apparently reverted to the lion. Coronelli showed two yachts in the 1690s, apparently very similar except for the heads. He shows the exterior of one with an equestrian figure. The other, with an interior view believed to be of the *Isabella* of 1683, has a figure of Neptune. The *William and Mary* apparently had a pair of cherubs side by side representing the dual monarchy, with another pair behind. The *Peregrine Galley*, originally planned as a small warship, had a rather upright lion as shown in paintings and a model. The new *Royal Caroline* of 1749 had the most elaborate of all, an allegorical scene with cherubs surrounding a central figure wearing a crown.

The rail of the head might be highly decorated in Stuart times, to such an extent that it almost ceased to be a rail. Witsen wrote of the *Katherine*: 'All round on the outside for two feet below the rails it is covered with costly sculptures and gold-painted carvings of grotesque figures and plants.'[10]

The stern

No yacht had an open gallery at the stern with balustrades, as was common on larger warships until the 1790s. It would be difficult to fit one on such a small vessel and, in any case, the distinguished passengers had plenty of room to promenade on the decks.

Witsen described the *Katherine*'s stern:

> At the top it boasts three round gilded lanterns; a maiden in stately dress, representing England as it seems, finely carved and gilded stands against it ... bearing in her raised right hand the arms of his Majesty; in her left hand she holds a sceptre with a lily on it; this hand holds down a stooping monster; the rest of the stern below is decorated with birds and other fine carved work, and the whole is covered with a royal crown.

The stern of a Stuart royal yacht was on four levels. The lowest was an area known as the transom, flat on a yacht and unlike larger English vessels, on which it was rounded. It was little decorated, though the Coronelli yacht shows a garter star on each side of the rudder. Above that was the counter, which usually curved inwards. In the first batch of Stuart yachts it was dominated by the mullions stretching downwards from between the windows above. Later it had no carvings but was decorated with a painted frieze.

The main feature which controlled the layout of the stern above that was the number of windows. The *Charles*, *Cleveland* and *Katherine* are all depicted with two large ones, as was the Coronelli yacht. Around 1710 the *Caroline*, *Fubbs*, *William and Mary* and *Mary* are all shown with four windows, with slightly different spacing. Whether there were two windows or four, there was usually a gap in the centre, decorated with a carved figure or a royal coat of arms. It is particularly prominent in a Van de Velde drawing of the *Charles* yacht of 1662. The windows were usually separated by carved mullions, straight or curved.

At the highest level, under the taffrail which formed the highest level of the stern, was the most heavily carved part of the ship. It usually had a royal coat of arms as the centrepiece, sometimes prominently supported with a lion and unicorn, and perhaps with cherub figures, which were popular at the time. The Coronelli yacht had a variant of this, with the coat of arms replacing the figure between the windows and equestrian figures in battle above. A model which is tentatively identified as the *Charles* of 1675 has the cross of St George inside the belt and motto of the Order of the Garter inside a star as centrepiece, with rather ugly cherubs on each side.[11]

The stern of the Coronelli yacht showing the arms of William and Mary and typical allegorical decorations of the age. The lines of the hull are also depicted. (From *Ships and Vessels*)

The gilded figurehead of the *Royal George* of 1817, depicting King George III. He is wearing a laurel wreath as victor of the Napoleonic Wars. (© National Maritime Museum, Greenwich, London, FHD0099)

The stern of the *William and Mary* of 1807, showing a plainer style. (© National Maritime Museum, Greenwich, London, NPD0666)

The stern carving of a model, probably of the *Charles* of 1675. (© National Maritime Museum, Greenwich, London, SLR0379)

The stern of the new *Royal Caroline* had:

> Taffrail and quarter pieces and case under the rich figure in midships, turned under the lights, counter rail and frieze between, figures and ornaments to the counter; brackets under the quarterpieces and lower counter rail; carved badges on the back and sides of the rudder, stern sash mouldings and spandals over ditto, port and top lanterns and carved stools to ditto cranks and bases, trucks and flagstaves and yard irons fore and aft.[12]

ART AND DECORATION   135

Sides

The corners of the stern were formed by carved quarter-pieces, often large warrior figures. The first features to be seen on the sides were rectangular quarter lights, windows which gave light to the main cabin. They were decorated round the sides.

The sides themselves were dominated by the gunports, with real ones towards the bows and false ones in the area of the royal quarters, which could not be interrupted by guns. The ports were often round and usually quite small, just large enough to accommodate the gun barrel with a little clearance, for they were only used for saluting and there was no need to elevate and traverse them as in a warship. Round wreaths were becoming standard for gunports by the time the Stuarts were restored, and that remained common throughout the eighteenth century. The Coronelli yacht has the stars of the Order of the Garter painted or carved on the sides, and a star on each side of the lower counter, a rare use of decoration in that space.

With the gradual restriction of carved work, there was a tendency to embellish the spaces between the gunports with allegorical paintings or gildings. In 1742 the Deptford officers reported:

> The carved works of his Majesty's ship the *Royal Caroline* being repaired in many places, and having new gunwales made in the waist which will require to be gilt, and as the pendulum frieze and frieze between the sheer and waist rail between the fore and main channels and some of the figures on the quarters are very much sullied, and will require some gilding also …[13]

The bulkheads on deck were usually decorated externally, though not as extensively as the stern. In the case of the *Royal Sovereign*:

> The fore and aft bulkhead to have double framing with mahogany, the lower panels raised, the upper panels lined with damask to ship and unship on frames with piliction mouldings, stanchions and ?? for the door; the door the same as for the bulkhead, the upper part of the door prepared for glass, if ordered the foremost bulkhead of the water closet to be double, inside framing of mahogany, the beams, upper panels the same as the fore and aft bulkhead outside framing deal with raised panels, the ship's side and roofing to be performed the same as the cabins, the front of the seat framed as the bulkhead, top clamped, cistern properly fixed …

Gilding

Gilding of the carved works using gold leaf was expensive and was often a bone of contention, though it was regarded as essential, especially for the principal yachts. In 1723 Henry Turner gilded the *Royal Caroline* at a cost of £75, but the Deptford officers reported:

> After he had finished it, there appeared a necessity of new gilding a great deal more of the old carved work than was at first proposed, it looking a great deal worse when it came to stand by the new work than was apprehended when it was seen by itself; he has expended 30 hundred leaves of gold in these additional works …[14]

The problem recurred in 1737, even though Turner had used 189,050 leaves of gold at a cost of £1417 17s 6d in 1734.

> Having caused the carved works of his Majesty' ship *Royal Carolina* to be new gilt, amounting to our estimate to £300, … we pray leave to acquaint your honours (that unless we leave the said ship imperfectly done) there will be a necessity for laying out about £200 to complete the gilding of the carved works and lanterns and new gild her

The sides, showing the Garter stars which were also used on other yachts. (From *Ships and Vessels*)

boat; we finding it impossible to make the ship appear uniform and free from patches but by laying on more gold than we first intended …[15]

Internal carvings

The only surviving carving which is apparently from a Stuart royal yacht is a coat of arms of James, Duke of York, which probably decorated the cabin of his yacht *Anne*. It is placed on top of an anchor, to establish its maritime role. It is made in lime wood, which is characteristic of Grinling Gibbons, and has the three-dimensional and vivacious style which is associated with his work.[16] The yacht in the Coronelli print has relatively restrained carving in the dining cabin, though it is difficult to see what is carving and what is painting. It is more elaborate in the royal bedchamber, with garlands behind the bed, which might or might not be from the Gibbons workshop, and more decorations round the bed canopy and around the windows.

The new *Royal Caroline* was perhaps the most elaborate royal yacht so far. As described some years later, her 'after room or bedroom' had mahogany window boards, bulkhead frames and panels. The nautical features, the rudder head and mizzen mast, were concealed behind carvings. Forward of that the awning room had looking glass doors on the bulkheads, damask panels and architraves round the doors. The stateroom had 'capitals to the pilasters, key stones, carved mouldings round the beams and damask panels, plain mouldings and beads under the beams and round the arches, deep carved panels supporting the handrails of the staircases leading from the state room to the after room and quarterdeck, festoons and the ornaments round the mizzen mast.' There were 'painted panels, consisting of history, battle pieces, shipping round the state room … and the panels in the shambles or companion representing trade and commerce'.

Paintings

The Van de Veldes, father and son, were lured from Holland in 1674 and were the greatest marine artists of the age, and perhaps of all time. In 1677 King Charles ordered Phineas Pett III, who was building the *Charlotte* at Woolwich, to arrange with Mr 'Vandevell' of Greenwich to produce paintings on panels showing 'the posture of ships in several ways' for installation on board. Two of these have been identified in the collection of the National Maritime Museum. One shows *Charles Galley* sailing from left to right, the other the *Woolwich* going in the opposite direction. It seems that they were put on opposite sides and both heading forward in the direction of the yacht. Another of similar size is dated 1675 and shows the *Portsmouth* yacht. It might have been intended for the yacht of that name.[17]

Their location on board is suggested by the Coronelli print, which shows four paintings on one side of the central dining room. Each depicts two or three square-rigged ships heading in various directions.

The Admiralty ordered:

> … that the paintings provided by Mr Pett, the master-builder at Woolwich, by his Majesty's own command (which he was pleased to declare the whole to be) as they are now to be found in and upon his new yacht the *Charlotte*, be with all speed satisfied by and for the officers of the navy, according to such agreement (if any) as Mr Pett did make for the same, or for such reasonable rates as the said officers of the navy can now agree with the painter for.[18]

The masterpiece of the Turner workshop at Deptford was the decoration of the new yacht *Royal Charlotte* in 1734. It produced 'the panels of the state room and the panels of the outside of the awning room of the said yacht with several land and sea pieces in the manner of basso relievo'. There were several paintings in the cabins, including landscapes, 'a piece representing a yacht and a smack in a calm' and '[t]he *Royal Caroline* and *Mary* yachts in a calm'. They were not all maritime: one showed 'a waterfall and a man drinking of water', another had 'a sutler's booth with soldiers going to refresh themselves'. The officers of Deptford were impressed and remarked on 'the goodness of these extraordinary paintings'. Nothing of them appears to have survived the yacht, which was rebuilt in 1749 and lost at sea in 1762. We can only speculate on who painted them. Was it Turner himself, or did he employ someone? Charles Brooking, arguably the finest marine artist of the century, was said to have been 'bred in some department of the Dockyard at Deptford', though that is now disputed and, in any case, he was only ten at the time. His father was a painter in Greenwich Hospital nearby, which is one possibility. The other marine painter of the area and period, John Cleveley, was around twenty-two by this time. He was not yet working in the dockyard, but in 1750 he was employed by the contractor John Stock do paintings for the *Royal Caroline*, so it is quite possible that he was also employed by Turner in this case.[19] He was paid for '[t]wo new large panels of shipping painted in basso relievo heightened in gold by Mr Cleveley at £100 each' and '[o]ne old panel in the front of the beaufit [?] new painted by Mr Cleveley' for £3.[20]

It might seem strange to include guns under art and decoration, but those of the royal yachts had little or no military function, being intended to return salutes fired by ships and fortresses in honour of royal and other distinguished passengers. That, however, was probably not the case with the *Isabella* in 1697, for the yachts had served as warships during the recent conflict.

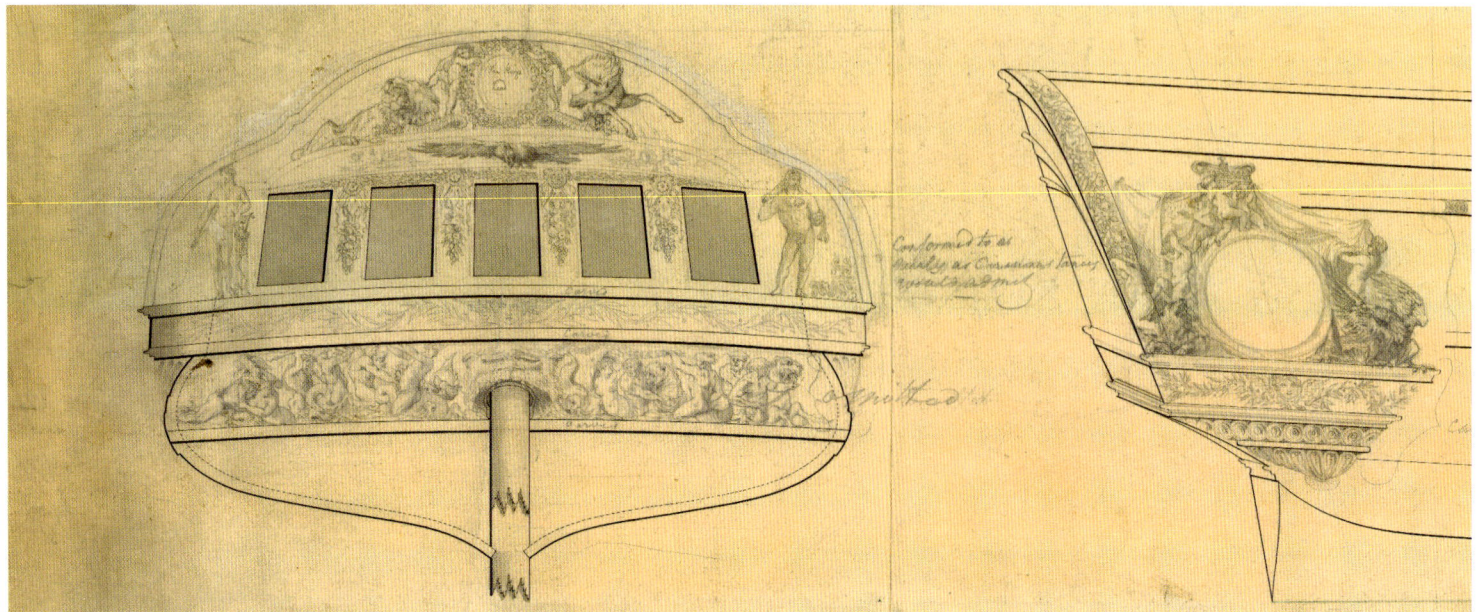

Decorations proposed for the *Royal George* of 1817, largely by painting rather than carving. (© National Maritime Museum, Greenwich, London, NPD0768)

The nineteenth century

The late Georgian yachts were generally more restrained in their decoration. This fitted in with the policy of the navy, which was ordered to 'explode carved works' in 1795. Yachts, of course, were exceptions to most rules, but there was a new-found restraint in most cases.

According to Lieutenant John Boteler in 1822, the *Royal Sovereign* of 1804 'was a beautiful ship, richly gilt, with a family head, as it is termed, the King, Queen and two or three children in a group. Her ports were all circular, with carved figures round them, of the size of a two-year-old child: her stern was covered with figures.'[21] According to a press report of 1822, the *Royal George* was 'a noble vessel, with an exterior distinguished for its symmetry rather than for its embellishments' and her quarterdeck 'scarcely exceeded, in decoration, one of our crack frigates', which the *Evening Mail* seemed to think was laudable.[22] There was a plan to decorate her with elaborate paintings – a classical family scene on the quarter, heads wearing laurel wreaths between the gunports and women, or goddesses, playing musical instruments on the bows. This was never executed and, apart from the usual figurehead and stern carvings, she only had carved dolphins on either side of the gunports with a plain lintel and sill above and below.[23] The yacht was described in the press in August 1822. Her exterior was 'distinguished for its symmetry rather than for its embellishments', and on board the same 'elegant simplicity was noticeable', despite a gilt coat of arms at the bows and a gilt tiller.

The stern of the *William and Mary* of 1807 had five windows like a small frigate, with rather plain mullions between. The counter below was undecorated, but above there was a royal coat of arms with figures of gods and devices curving round the quarters. The *Royal George* of 1817 also had five windows, but the decoration was mainly by painting. The mullions had floral devices and there was a painted goddess and a giant on either side; above there was a wreath supported by cherubs, and by a lion and unicorn in unusual posture. A complicated frieze on the counter was probably not executed.

The figurehead of the *Royal Charlotte* of 1824 survives. It is apparently carved in pine and is 2.159m deep, depicting the young queen as she was around sixty years earlier, perhaps not long after her traumatic voyage from Germany. As described it shows her:

> … at just over half length, crowned, with an orb held before her in her left hand and a sceptre … in her right, trailing slightly behind the figure. To either side she is supported behind by attendant cherubs with, on her right side, the Hanoverian royal arms, and on her left the union flag, both within shields surrounded by foliate carving, and with flowers on the centreline where the figure slots over the supporting stemhead.[24]

But the fashion for expensive decoration was clearly over. Though the growing Victorian middle classes loved to embellish their homes inside and outside, their representatives in government tended to be critical of government expenditure, and the steam yachts which were to follow would be relatively plain. The era of Stuart extravagance was long gone.

5
Sailing the Yachts

Fore-and-aft rig and square rig

All the early royal yachts were fore-and-aft rigged: that is, their more important sails were set fore and aft in their neutral position. As such, they were best as sailing close to the wind, whereas square rigged ships were best with the wind behind. The two-masted, square rigged ketch was introduced with the *Fubbs* and *Isabella* in the 1680s. Square rig begin to dominate among the yachts when the *Peregrine Galley* was converted and renamed in 1716 – though she had served as a yacht before then. After that the principal yacht, the one normally used by the King, was usually square rigged. But in the early days the majority of yacht captains and crews had to be familiar with fore-and-aft rig, which did not come naturally to the navy. All major warships were square rigged. Single-masted vessels were usually known as sloops, but the most successful sub-type, the cutter, did not enter the regular navy until after 1763. Manuals of seamanship appeared rather late in the day and offered little guidance on fore-and-aft rig. Falconer's comprehensive *Marine Dictionary* said frankly: 'The design of this work being professedly to treat of the construction, mechanism, furniture, movements and military operations of a ship'[1] – that is, a vessel with three masts and square rig on each, like the larger yachts. The most detailed sailing manual of the age, D'Arcy Lever's *Young Sea Officer's Sheet Anchor* of 1819, said nothing about fore-and-aft rig. An officer appointed to the Stuart yachts would probably have learned his trade in much larger ships with different sails and he would have to master a new kind of seamanship, though he might be assisted by a master with experience in small merchant ships.

The pattern for the original single-masted rig was set by the *Mary*, which had one 'mast of good quality', worth 100 Dutch florins (Fl), a topmast worth 12, a gaff with peak and jaws valued at Fl 22 and a big and a small boom worth 10 and 6 florins respectively. There was a bowsprit and knee costing Fl 8, a 'wide foreyard' at Fl 5 and a topsail yard for Fl 4.[2] As this makes clear, the Stuart yachts, apart from the Bezans, were not entirely fore-and-aft rigged. Each had a yard set at right angles to the mast, with the capacity to hang a square sail below it, and a smaller yard above that for a square topsail. There is no sign that the crew climbed up the mast to furl the sails as they did on square-rigged ships. The topsail could probably be hauled up; the *Mary* inventory mentions '1 pair of clew lines for the top sail', which could be used to haul it, but often they were removed and stowed when not in use, and the yards were canted or angled to keep them out of the way.

The yacht *Cleveland* in a squall with a ship of the line and another royal yacht in the distance. She is on the starboard tack in a rough sea, and the gunports look dangerously exposed. Both the topsail and jib are flogging, which suggests the sheets have been eased, or that she's been pointed up a little too close to the wind. (Private collection, taken from Gavin, C M, *Royal Yachts*)

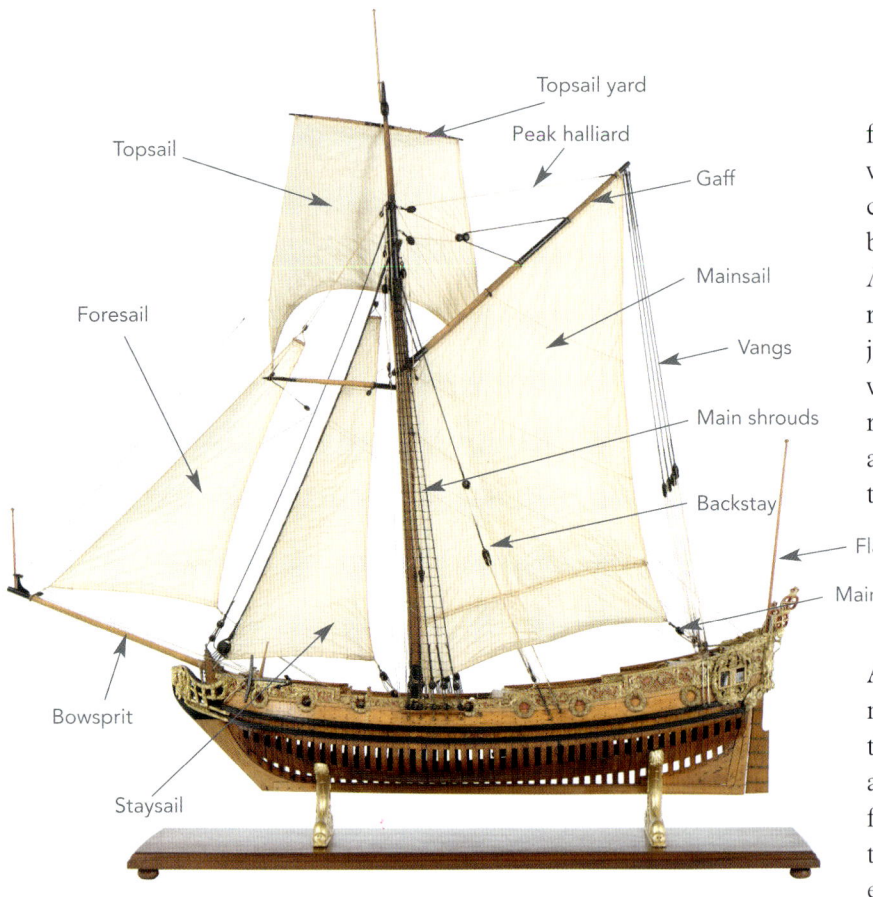

A model of a yacht of around 1690 showing the standard single-masted rig. (© National Maritime Museum, Greenwich, London, SLR0378)

∽ Ketch rig ∽

Pepys credited Charles with introducing the ketch rig to yachts, mentioning 'the King's present invention of applying two masts to a yacht'.[3] It was first used on the *Fubbs* and *Isabella*. Ketch is defined as a two-masted rig in which the forward mast is shorter than the after, as distinct from a schooner which became common later, and in which the masts are the other way round – therefore a ketch has a mizzen and a mainmast. To Pepys and his contemporaries, it did not imply fore-and-aft rig. The bomb ketch, which was developed by the French and then the English in the 1680s and 1690s, was essentially a ship with the foremast missing, to allow the firing of mortars. The ketch-rigged yachts had square sails on the main and a square mizzen topsail; the lower mizzen sail was lateen rigged with a triangular sail under a diagonal yard, and later gaff rigged with no part of the yard or sail forward of the mast. The mainmast was still placed near the centre of the ship with a smaller mizzen aft. This could have been very unbalanced, but was partly compensated by a long bowsprit carrying a very large jib and staysails forward of the mainmast. When sailing with the wind behind these had little effect and the principal effort came from the main and mizzen, which could make the vessel more difficult to steer. And it was not particularly good at sailing into the wind or for tacking.

Nevertheless, the ketch rig gradually became common for yachts. After the *Fubbs* and *Isabella*, the *William and Mary* was built with it, but not without difficulty. According to the captain's log of November 1694, 'took out the mizzen mast, it being too short ... I set new mizzen mast and rigged it.'[4] The *Mary* was converted to ketch rig in 1735. But already the ship-rigged *Royal Caroline* was the principal royal yacht and she was joined by the *Augusta* in 1771. The nineteenth-century yachts were all ship rigged. This meant at least three masts with square rig on all of them, though the lower mizzen sail would be fore-and-aft, and staysails were rigged between the masts and from the bowsprit to the foremast.

∽ At anchor ∽

Anchors were particularly important for the royal yachts. They needed them when wind and tide were unfavourable during the passage up and down the Thames, and they often had to anchor off ports such as Margate and Hellevoetsluis to await a fair wind, or the arrival of distinguished passengers. In 1737 the *Fubbs*, *William and Mary*, *Katherine*, *Mary* and *Charlotte* each had five anchors – two of 9 or 10 hundredweight (cwt) each, known as the bower anchors because they were usually hung from the bows; a slightly smaller one, the sheet anchor, of 8 or 9 tons, a kedge anchor of around 4 tons, which could be rowed out ahead under a boat and the ship hauled up to it; and a smaller stream anchor with a similar function. The captain of the *Katherine* wanted his sheet anchor increased from 8 to 10cwt, as the existing one was 'too light for the yacht's use'. In 1794 the *Princess Augusta* had a best bower, a spare anchor only slightly lighter, a small bower, a stream anchor, a 'cage anchor' and a grappling anchor. As fitted in 1802, the *Royal Sovereign* had three bower anchors of 13, 14 and 16cwt plus a lighter stream anchor.[5]

Captains sometimes complained that they were too small – for example, for the *Isabella* in 1706, the *Fubbs* in 1731 and again in 1737 and the *Katherine* in the same year. The *Fubbs* was almost lost when her cable parted off Harwich in 1731. Often anchors were lost, for example by the *Henrietta* in a gale in 1702 and the *Fubbs* at the Nore in 1731. They might be recovered, such as that of the *Isabella* in Margate Roads in 1712 by John Grant– anchor recovery was a regular source of income for the fishermen there and at Deal, where a Mr Benjamin 'took up an anchor cable left by the *William and Mary* yacht off North Foreland' in 1766.

A voyage usually began by loosing the ropes which held the yacht to a buoy off Greenwich, or by raising the anchor. This was done by means of a windlass in most yachts, as it took up less space than a capstan, but required more skill. The *Royal Caroline* had a capstan, perhaps because of her origin as a sloop of war; but her successors, even if larger, had windlasses.

Placed well forward, they did not interrupt the midship area where the distinguished passengers might promenade, and it was perhaps assumed that the yachts had enough skilled men to handle a windlass. The anchors were hung under catheads in the bows, and the rope cable was stowed in the hold.

The basics of sailing

The royal yachts were usually steered by means of an iron tiller. There was no practicable alternative in Stuart times, except a rather clumsy arrangement known as the whipstaff, which was only used on larger and multi-decked ships. The steering wheel was invented around 1700, but the system needed a certain amount of slack in the ropes until improvements in the 1770s. It was not applied to smaller vessels such as the yachts until after that – *Royal Sovereign* of 1804, *William and Mary* of 1807, *Royal George* of 1817 all had wheels right aft. The *Prince Regent* of 1820 had a peculiar arrangement involving a geared quadrant and crank.

William Falconer identified four conditions of sailing – close-hauled, or into the wind as far as possible; large, with the wind coming over the beam and allowing a good deal of freedom of movement; quartering with the wind about 45 degrees aft of the beam and the best point of sailing for many vessels; and afore the wind, which had advantages and disadvantages. Close hauled was where the fore-and-aft rig had the advantage. Falconer, referring mainly to square rigged vessels as usual, wrote, 'In this manner of sailing the keel commonly makes an angle of six points [135 degrees] with the line of the wind, but sloops, and some other small vessels, are said to sail almost a point [22½ degrees] nearer.' However, all vessels would make about a point of leeway as they were pushed back by the wind, and more in strong winds. It was pointless to try to sail against both wind and tide. Sailing close-hauled literally meant that the sheets which controlled the corners of the sails were hauled in as tight as possible. They could be hauled closer in fore-and-aft rig, hence the advantage. The evidence of drawings and paintings suggests that the captains of the Stuart yachts often had the square topsail set even when sailing close to the wind. It required careful helmsmanship to get the best out of the ship in these conditions. The helmsman was ordered to steer 'full and by', 'sailing in such a manner as neither to steer to nigh to the direction of the wind, not to deviate to leeward; both of which movements are unfavourable to her course'.[6]

If it was nevertheless necessary to try to make progress directly into the wind, the process of tacking was used, with the head of the ship being turned into the wind to alter course through at least 135 degrees. Yachts had the advantage that their smaller hulls would allow them to turn more quickly. A square rigger, however, could back the sails on its foremast to help the ship turn through the wind, while a sloop or ketch had only the less effective force produced by backing the jibs. It is comparatively rare to find yachts tacking, because mostly they waited for a fair wind. However, Van de Velde shows such a manoeuvre off Sheerness in July 1673.[7] In 1709 the *Mary* tacked inshore to find suitable ground to anchor in. In 1729 the yachts were anxious to get into Hellevoetsluis to meet the King and the log of the *Mary* records that she 'plied to windward' in 'several short trips', anchoring when the tide was against them. But they failed in their object, for the King had already sailed by the time they got there.

Tacking could also mean turning into wind to bring the wind on the other side. The procedure for tacking was described by Joseph Goodall in the *Princess Augusta* in 1795. The order 'Hard a lea' was given, then:

> Let go the fore sheet, and stay sail sheet, check the fore top bowline, raise the fore tacks, haul the main topsail. Brace about the head yards, or let go and haul when the ship comes to, brace up and haul aft, then haul the bowlines, haul tight the weather braces lea tacks and weather sheets in their beckets.

The alternative was veering or wearing, by turning the stern towards the wind. This was not often deployed by yachts; it needed a good deal of space, so it was not viable in river passages, and was more suited to square rigged ships.

According to Captain William Parry in the *William and Mary* in 1746, she was not happy in quartering. 'The wind abaft the beam brings the sea upon her quarter, she then steers wild'. It was better when the wind was slightly ahead: 'when two points before the beam she steers easy and so in proportion as the wind veers ahead'. Speed varied with the conditions but was generally satisfactory: 'abaft the beam in fresh gales she runs eight or nine knots; with the wind on the beam and a fresh gale eight knots; when close hauled on a bowline, smooth water, seven knots; she carries her helm in the above mentioned pretty much a-weather.'

When sailing 'afore the wind', the earlier fore-and-aft rigged yachts had the use of the square main and topsails. In 1673 Van de Velde showed them with the fore and aft gaff sail still set, but not catching the wind. Sailing 'afore the wind' without the square sails, as in the later yachts, also had its problems. The gaff mainsail was let out as far as possible but it might mask the other sails set ahead of it, losing power – there is no evidence that the yachts sailed 'goose-winged' as later generations would put it, with the headsails set on the opposite side. Sailing before the wind also needed careful steering to prevent the mainsail coming over, while the wind induced a rolling motion.

For setting and reducing sail, the *Fubbs* was old-fashioned in 1719 in using a 'bonnet', an extra strip of canvas fitted under a sail to increase the area, in this case of the jib. But she also carried the means to reef the mainsail. This was described by Falconer as 'the operation of reducing a sail, by taking in one or more of the reefs … The intention of the reef is to reduce the surface of the sail in proportion to the increase of the wind; for which reason there are several reefs parallel to each other in the superior sails.' In September 1712 the log of the *Peregrine Galley* records, 'blowing fresh, took a reef in each topsail.'[8] The rebuilt *Royal Caroline* of 1749 evidently had two rows of reef points on her topsails. The *Augusta* of 1771 had three rows on her topsails, as did the *Royal George* of 1817, with a single row on the lower masts.[9]

Sailing in rivers

Almost all the royal yacht voyages began with a passage down the winding waters of the Thames, even if the distinguished passengers chose to join the ship at more convenient points. The trip needed favourable conditions and ships would almost invariably have to anchor at times to wait for a favourable tide. A large ship like a 74 or frigate would be very difficult to tack, or zigzag into the wind, in the narrow and crowded river. A fore-and-aft rigged vessel, like the smaller yachts, could sail a point closer to the wind, took up less space in tacking and was quicker to handle. With a favourable wind it was possible to use the tides for a yacht to get downriver in a few hours. In 1749 the Board of Admiralty was received on board the *William and Mary* off Greenwich at three in the afternoon of 2 July, carried out some salutes and ceremonial, and reached the Nore at the mouth of the river by 7.30.[10] But the position was very different when the winds were light or unsuitable.

Om 24 January 1726 the *Mary* slipped her mooring at Greenwich and was towed down to Erith in very slight winds, presumably by her boats rowing ahead. After eight hours at anchor, she proceeded down to Gravesend and anchored again, in 'little wind and fair weather'. A small breeze sprang up in the evening and she was able to progress and anchor 'near the buoy of the Middle'. There was thick fog next morning but they sailed after it cleared around ten and 'plied down to the Nore, and kept plying till low water then anchored in seven fathom water'. They weighed again at high water at eight in the morning of the 27th and reached the Swin, but 'At noon the tide being done I anchored near the Shoe Beacon.' They lay there all night and proceeded to the buoy of the Spitt next morning but anchored again when they met the flood tide. They reached the Gunfleet off the coast of Essex and after another night at anchor they found a fair wind to take them across to Holland.[11]

The yachts had the advantage of shallow draught. In November 1725 Admiral Norris wrote to the captains of the escorting ships from the *William and Mary*, 'As I intend to go over the flats with the yachts, and you cannot keep me

A Van de Velde drawing showing the yachts heading down the Thames for a visit to the *Tiger* in 1681. Those on the left – the *Mary*, *Cleveland* and *Charlotte* – are sailing on a broad reach. The next two, the *Kitchen* and *Henrietta*, have turned to starboard and are now running before the wind. The *Isabella* is to the right. (© National Maritime Museum, Greenwich, London, PAH1870)

company in that passage, you are with the first opportunity of wind and weather to proceed down the King's Channel to our rendezvous at Hellevoetsluis.'[12]

The yacht officers were familiar with the numerous sandbanks on the western side of the North Sea, incidents like the loss of the *Gloucester* excepted, but it was a different story in Dutch waters, where they often had to employ a pilot, with varying degrees of success. Information from a fisherman during William of Orange's first return to Holland almost led to disaster. And in 1732 it was reported:

> When the [*Royal Carolina*] had got within reach of a Dutch pilot, one of that nation came on board, and … he answered he could not promise to land his Majesty under to tide's work …; at which the King, being somewhat uneasy, sent again to consult Goodwin [the master] who still persisted, notwithstanding the Dutch pilot's dilatory opinion, that he should certainly (with God's assistance) land his majesty in the time he first prefixed. Thus it exactly happened, as all the watches witnessed.

It was more serious in 1704 when another Dutch pilot ran the *Mary* aground just after high water, the worst possible time. The falling tide would make it impossible to get off at least until the next high water more than twelve hours later. In fact, it would take much longer. The *Mary* unloaded guns, spare anchors and cables into Dutch barges or schuyts, but still she did not float, in fine gales and a hard frost. After four days, 'This day we tried to lift her at high water with two great schuyts but could not, the tide not coming in enough.' On the next day, 'at four in the afternoon we floated and hove off to our anchor and took on board most of our things again.' Going up the River Scheldt to Antwerp in the *Royal Sovereign* in 1822, Lieutenant James Boteler recorded, 'We worked up the river,

A ketch-rigged yacht off Dover, c1760. The square main course is set and hauled round as far as possible, obscuring the view of the other sails. (© National Maritime Museum, Greenwich, London, BHC1069

which winds immensely and, in parts, quite turns back. We had much work; bracing-up on one tack, then perhaps on another, then before the wind, and up or down studding sails. I never before or since had so to knock a ship about'.[13]

∽ Bad weather ∽

The smaller yachts were particularly vulnerable to bad weather. In 1775 Joseph Banks reported from the *Augusta*, 'Night dirty and blowing, topgallant yards down, much pitching and flunking and seasickness, in short a perfect yacht-storm.'

Captain Molloy of the *Mary* was off the coast of Kent in October 1729 on the way to pick up the Duke and Duchess of Richmond at Calais. The weather seemed 'clear and favourable' as he left his anchorage off Margate at midnight on the 19th, but as he rounded North Foreland he encountered 'fresh gales and squalls with rain' from the south-southwest. He took in the topsail and stood away to the southeast with the lower sails, but by eight in the morning the gales were increasing with 'a great tumbling sea'. He brought the yacht head to wind with only the peak of the mainsail set to control her and at four in the afternoon he had four of the guns lowered down to the hold to increase stability – only possible with the very light guns carried by yachts. He took precautions to prevent water entering – 'nailed down the paulings [tarpaulins?], and shut up the deadlights of the state room windows.' At ten in the evening the storm abated, he was able to set the topsails at eight next morning, but he was dangerously close to the Flanders shore at Blankenberge and the lead had to be heaved to check the depth. He was well east of his destination of Calais but was able to ply to windward to reach it in time.[14]

∽ Sailing trials ∽

In February 1748/9 the yachts *William and Mary* and *Fubbs* were ordered to be cleaned and tallowed alongside the new *Royal Caroline*, and Captain Molloy was:

> to proceed with them and his Majesty's ship under your command to sea and cruise down the Channel as far as you shall judge fit in order to make a trial of her sailing in comparison with the other yachts, and when you have made a sufficient experiment of her sailing you are to return with them to their usual moorings in the River of Thames and make an exact report to us of your proceedings.[15]

The three yachts were docked, cleaned and tallowed to make the test fair.

Molloy reported, 'I sailed with the two yachts in my company from Deptford. We made a plain sail and differed only as the windings of the river obliged us to alter our sails, when we were upon one mast the yachts held us way, but in those reaches where our head sails drew we wronged them much.' This presumably means that with the wind directly behind only the sails on one mast would fill, which was a higher proportion of the motive power of the smaller yachts; with the wind on the beam the sails on all three masts of the *Charlotte* would fill, as well as the headsails forward of the foremast, giving her an advantage.

Out of the river, they yachts spent a few days sailing just outside the estuary, from the Downs to the Gunfleet:

> Wednesday morning I weighed, the wind being aft and stretched away to the north east four or five leagues with a fine moderate gale and smooth water and had a short trial with the two yachts being close hauled upon a wind and notwithstanding the moderation of the weather which was much in their favour, considerably wronged them, having in the space of three hours forereached and weathered them at least two miles. After we had got well to the eastward we bore up and went the back of the Goodwins for the North Foreland run, though the yachts crowded with their topsail studding sails which we spared them, they both dropped pretty much astern.

This means that the *Caroline* sailed well close to the wind, though it was light, and equally well with the wind astern, though the yachts used extensions to their sails, which the *Charlotte* did not need. The next day was less productive. 'Next morning, Thursday, I weighed with a fresh gale at north-west and stood to the eastward about four or five leagues, but the wind dying away we had no great trial to speak of that day.' Later, Molloy went on:

> having as much wind as I could carry my topsails double reefed which gave us an opportunity of a fair trial with the yachts which we much coveted and hoped for, we fairly made sailing together and within the compass of four hours I distanced the *William and Mary* at least a league right in the wind's eye, and for the *Fubbs* almost ran her out of sight …

So the *Caroline* was equally good in strong winds.[16]

~ A race ~

Having sailed with Cook during his three-year circumnavigation of 1768–71, the naturalist Joseph Banks had had plenty of opportunity to observe seamanship as executed by one of its greatest masters. He was with Sandwich in the *Princess Augusta* during his 1775 trip to inspect the Royal Dockyards, when the officers of the *Mary* 'boasted so much of the going of their vessel that, (loth to give up the honour of our own) we agreed to run with him the next morn, which were got first into Dover Roads to win.'

At first the *Augusta* did well with the wind behind. 'Before nine weighed and sailed down the Queen's Channel with a large wind, the *Mary* following us, found that we went ahead of her considerably.' But things changed after they emerged from the estuary:

> At 11 passed the eastern buoy of Margate Sand and hauled close upon the wind in order to pass around the Foreland, found that in this situation the *Mary* went to windward of us so much that though in the evening the wind came large again and we gained upon her very much, yet she anchored off Dover about a cable's length ahead of us and consequently won.

The *Augusta*'s officers and passengers went on board their rival, and 'so little did we give up our superiority that we matched the yachts once more to sail to Spithead'. Consequently, '[s]tarted this morn with the first of the tide, which happened very early in the morn; unfortunately for us the wind very foul so that at breakfast the *Mary* was very considerably to windward of us.' By the evening both yachts were becalmed and the *Augusta* anchored a league astern of the *Mary* to prevent being driven backwards by the tide. A small breeze got up after half an hour and both ships used sweeps and boats' oars to make slow progress, and Banks claimed that they 'neared the *Mary* considerably'. They passed Brighton next morning with a foul westerly wind and the *Mary* 'very considerably ahead'. The *Augusta* reached St Helen's Road at the eastern end of the Isle of Wight that night, with the *Mary* anchored some way ahead at Spithead. Banks conceded, 'we were again beat, and rather shamefully.' In the morning they weighed and headed up the Solent but the *Mary* added insult to injury by coming back to meet them.

After Charles II's early days of racing, yachts did not need any great speed but captains inevitably regarded it as the best feature of their ships. Since their voyages were quite infrequent, they probably relied on the experience of their crews rather than practising together. But traditional seamanship would matter less during the next century.

Part VII

Last Days Under Sail

1

Victoria in Scotland

At five in the morning of 29 August 1842, Queen Victoria and her husband Prince Albert boarded one of the broad carriages of the Great Western Railway at Slough Station, just four miles from Windsor Castle. It was only in June that the Queen had discovered the virtues of the line and she wrote to her uncle, the King of the Belgians, from Buckingham Palace, 'We arrived here yesterday morning, having come by the railroad from Windsor, in half an hour, free from dust and crowd and heat, and I am quite charmed with it.'[1] This time they were to repeat Victoria's uncle's trip of twenty years ago, to visit Scotland in a royal yacht. From Paddington Station a road carriage took them to Woolwich, where they were to embark. It was downriver from Greenwich and that would shorten the sea passage, and it was less suitable for a public spectacle, which might have been welcomed by the Queen, who was tired of company, and the courtiers around her who worried about security – there had already been two assassination attempts on her, most recently in May. Instead, the scene at the dockyard was dominated by naval and military personnel including artillerymen, engineers and marines, but a 'heavy, dingy-looking coal-barge had to be cleared out of the way.'[2] The couple went down waterside stairs to the royal barge, which was steered by Captain Sir Francis Collier. They boarded the *Royal George*, the same ship that had carried George in 1822. The Queen recorded, 'I write here, comfortable established with dear Albert in the very pretty cabin. How delightful to be quite alone together on board – no Ladies or Gentlemen, only our servants and of course Lord Adolphus [Fitzclarence], the captain, commander and the lieutenant and other officers.'[3] In fact the gentlemen were accommodated on the *Rhadamanthus* and *Shearwater*, and the ladies on the *Black Eagle*.

They started off with the yacht towed by the steamer *Monkey* (ex-*Royal Sovereign*, which had been bought for the navy in 1837) in company with the steamships *Shearwater*, *Salamander*, *Lightning*, *Black Eagle*, *Rhadamanthus* and *Fearless*. The *Salamander*, however, damaged her paddle box and had to stop for repairs, before catching up with the squadron at Tilbury – 'a miserable looking place', according to the Queen. They met other warships at the Nore, and it was found that the *Monkey* was having difficulty in towing the yacht, so the *Royal Eagle* took station ahead. The Queen enjoyed the flat waters of the first part of the trip. 'We walked up and down on deck, for some time, which was very pleasant, then, sat and read. There is a cow on deck, in a place specially arranged for her, where she seems quite happy.' But after lunch:

> There was some pitching, caused more by the wash of the steamers than anything else, and this continues. – Played a little on the piano. Sat, and walked about again, and then came down and rested. Was rather ill, but recovered again and went back on deck … Both of us feeling rather bad, were unable to take any dinner and retired to bed before 9.[4]

By this time the squadron was off the flat coast of East Anglia and passed through Hasborough Gat, eight miles off the coast of Norfolk – it was a feature of the voyage that they tended to sail close to the coast, which gave more shelter from the prevailing southwest winds, and was made easier by steam power and accurate charts, which the Queen perused during the voyage. In the morning the sea was calm again and the *Royal George* signalled to the other ships that the Queen and Prince were 'perfectly well'. The *Black Eagle* confirmed that the Duchess of Norfolk and Miss Paget were also well, while the *Rhadamanthus* replied, 'All well, and the Lord Steward eating monstrously.' They were off Scarborough by evening, but the wind freshened during the night, which delayed progress. By next afternoon they were approaching the Farne Islands, where Grace Darling had gained fame by rescuing the crew of the *Forfarshire*. True to form they sailed close inshore:

> This was, perhaps, one of the most interesting points of the whole voyage, the proximity of the shore on either hand, the rapid speed of the vessels as they swept past, hurried along by a strong ebb tide, the mingled roar of the cannon of Bamborough Castle, and the loyal cheers of the island fishermen who surrounded the squadron in

The *Trident* arrives back off Woolwich, with the Queen and Prince Albert on board. (© National Maritime Museum, Greenwich, London, PAH8905)

their boats, gave an animation to the scene that it is difficult to describe ...[5]

Off Northumberland the *Royal George* set her jib, fore staysail and flying jib to sail close to the wind, but continued to be towed by the two steamers. They passed Berwick and entered the Firth, which was impressively illuminated by beacons on the hilltops. They were behind schedule but at quarter to one in the morning of 1 September the Queen heard the anchor being dropped, 'a welcome sound'. They went on deck at seven and breakfasted. 'Leith was close on one side, the high hills over-towering Edinburgh, which was wrapped in fog; on the other side was May Island, where it is said Macduff held out against Macbeth; the Bass Rock was behind us.'

This time the royal party would not have to disembark by boat. The Duke of Buccleuch, their host on the trip, had built a pier from his land at Granton, which allowed ships to come alongside – the first part had been opened on Victoria's coronation day in 1838 and it was named after her:

> At 8 we arrived at Granton Pier, where stood the Duke of Buccleuch, Sir Robert Peel etc – they stepped on board, to see us, and the latter told us, the people were all in highest good humour, though naturally a little disappointed at being mailed for us yesterday. We then stepped across a sort of bridge on to the pier, the people cheering, and the Duke saying he begged to be allowed to 'welcome' us.[6]

This led to the most embarrassing incident of the trip. Unlike George IV, the Queen decided to come ashore as soon as possible and allowed little time for preparation. Protocol demanded that the City Council greet her and hand her the keys of the city, and they sent two Baillies, a councillor and a police official to Granton on Wednesday evening. Nothing had been heard of royal progress and it was assumed they could not arrive that evening, so the policeman was left to report. There was news at three in the morning that the squadron was in the Firth, but they only found out that it was off Inchkeith at six. After the *Royal George* came alongside: 'The Duke of Buccleuch and Sir Robert Peel then intimated to Ballie Richardson that her majesty found it necessary to proceed immediately to Dalkeith', and there was not time to alert the councillors or the public. Councillor McAulay went so far as to say that 'after the Queen did arrive at Granton Pier, her proper course was to remain on board until the Lord Provost and the rest of the Council had got their breakfasts.' This was

The Royal George approaching Granton Pier in 1842. The steamship is shown to the right, but like most pictures of the trip, its role tends to be understated. (Royal Collection Trust / © Her Majesty Queen Elizabeth II 2022, 919574)

greeted with cries of 'Oh, oh,' by his fellow councillors, but he went on to claim that there was 'universal disgust in the public mind'.[7]

Her Majesty did not accept responsibility. 'Though the crowds were really not so great, the crush was such that one was in constant dread of accidents. This need never have been, had there been more regularity and order; there had been some stupid mistake on the part of the Lord Provost about giving due notice of our approach.'[8] The incident showed that it was still very difficult to co-ordinate a long-distance royal visit, despite the advantages of steam power.

Nevertheless the Queen began to enjoy the trip:

> Edinburgh made a great impression upon us; it is quite beautiful and unlike anything I have seen. The town is so regular, everything built of massive stone (not a brick to be seen). The High Street, which is pretty steep, is very fine – then the Castle, situated on that perpendicular rock, in the centre of the town, is most striking, and Calton Hill, on the opposite side, with the national monument, a building, quite in Grecian style … with Arthur's Seat in the background, overtopping the whole. The enthusiasm was very great, and the people very friendly and kind.[9]

The Queen would travel a little more than her uncle, being introduced to the beauties of Perthshire. She did not say goodbye to water transport; 30ft eight-oared boats were built to transport her and the party fifteen miles along Loch Tay to Auchmore, the seat of the Marquess of Breadalbane. They were 'steered by the royal bargemen and rowed by sturdy Highlanders'.

The 1842 visit only had a limited success as a spectacle, and did not compare with that of 1822 in its general influence. Victoria, though a far more worthy person than her uncle George, paradoxically attracted more criticism, perhaps because the public now had much higher expectations. But the effect on Victoria herself was profound. On 10 September she wrote to Lord Melbourne her favourite politician, 'The Highlands and the mountains are too beautiful, and we *must* come back for longer another time.'[10]

The General Steam Navigation Company, a consortium formed in 1824, operated a variety of steamships in European waters, the Thames and the North Sea with a regular Leith to

London service. They had proposed that their new ship the *Trident* should be used to convey the Queen, but were told that 'Her Majesty … had been graciously pleaded to decline their offer, as arrangements had been made that her Majesty and Prince Albert should proceed in the Royal Yacht attended by several government steamers.'[11] But the Admiralty had anticipated that the practice of towing the royal yacht had its limitations and had the *Trident* fitted out to accommodate the royal party. The ship was 200ft long and of 1200 tons, nearly four times the size of the *Royal George*, which came from an earlier age. She usually did the passage in thirty-eight hours, about half the time of the royal voyage north. Taylor and Sons of Great Dover Street, Southwark, were employed to furnish her. The 30ft by 35ft saloon was fitted with a damask-covered sofa in the centre, and with ottomans round the sides and chairs covered with crimson velvet and edgings of gold lace, and a rich Brussels carpet on the deck. The mast passed through the saloon, but was covered with satin wood and rosewood to give the appearance of a flower. There was a library at the stern containing 'a choice supply of books well fitted to wile [sic] away time on the voyage.' The 12ft by 15ft royal cabin or stateroom contained two French beds and satin-covered chairs, footstools and a rosewood table. There were adjoining dressing rooms and five staterooms below for the royal suite round a central saloon.[12] The emphasis was on comfort rather than grandeur and display, as in the royal yachts.

It did not take much to persuade the Queen and Prince that this was a better way to travel. Victoria commented, 'On board the *Trident*, the accommodation is much more comfortable and roomy than on the *Royal George*, and she is beautifully fitted up.'[13] They left Dalkeith Palace, Buccleuch's house, where they had been staying due to renovation at Holyrood House. This time there was no mistake, a circular was issued by the Lord Provost making sure that 'the public were made fully acquainted with all the necessary details respecting her Majesty's embarkation', and large crowds thronged the route. They boarded the steamer at Granton Pier and soon the Queen recorded, 'As the fine shores of Scotland receded more and more, we felt quite sad, that this very pleasant and interesting tour was over, but we shall never forget it.' With the wind against them, they soon left all the other ships of the squadron behind, except the *Monarch* carrying the royal horses, which also belonged to General Steam Navigation. The Queen enjoyed her luncheon off St Abb's Head – 'a good sign at sea' – and read Sir Walter Scott's *Marmion* with its description of Holy Island, which they soon passed. They Queen and Prince argued about whether they would be able to see Alnwick Castle, but the Queen lost as it was in a dip. They played the piano and had dinner, but went to bed early as the wind rose. They arrived at Woolwich and were taken by carriage back to Paddington, then on to Windsor.

The Queen was converted to steam and wasted no time contacting the prime minister suggesting a new royal yacht and he replied through his secretary. 'Sir Robert Peel trusts that your Majesty may entirely depend upon being enabled to make any excursions your Majesty may resolve upon in the early part of next summer, in a steam vessel belonging to your Majesty, and suitable in every respect for your Majesty's accommodation.' Peel was aware that yet another nautical revolution was about to take place, with the adoption of the screw propeller. It had several advantages in a warship. It was not exposed to damage as paddles were, it did not restrict the broadside space available for guns, and it was far more comfortable in a rough sea when one paddle or the other might be lifted out of the water, making the ship difficult to control. The navy was already building the pioneering *Rattler* and Peel told the Queen, 'the Admiralty is now building a large vessel to be worked by steam power, applied by means of revolving screw instead of paddles.' He doubted if it would be as fast as a paddler, but he had been assured that 'the working of the engine is scarcely perceptible; that there is none of the tremulous motion which accompanies the beats of the paddles, and that it will be possible to apply an apparatus by means of which the smoke will be consumed, and the disagreeable smell in great measure prevented.'[14] But that was a step too far: the new yacht, the first *Victoria and Albert*, was launched in 1843 and propelled by paddles.

The railway revolution accelerated during the 1840s, from 1939 miles of track in 1842, to 6084 miles in 1850. By the end of the decade London was linked with Scotland by rail, with other lines going further north, and allowing the Queen and Prince to build their favourite home at Balmoral in 1853–56. Combined with the electric telegraph, the railway allowed royal visits to be organised with clockwork precision and there was no need for embarrassing incidents as in September 1842. But the new steam-powered royal yachts still had a role. The first task of the *Victoria and Albert* was to take the Queen and Prince on a tour along the south coast of England and over to France. Over the years it took them to places which were less accessible by railway including the Channel Islands, Scilly Islands and the west coast of Scotland, as well as foreign ports such as Antwerp, Ostend and Boulogne.

The sailing yacht was finished as a means of royal transport, but by the end of the century it would be revived for leisure purposes, arguably taking it back to its origins.

2
Second Wind

The steam yachts

No time was wasted in providing the first steam royal yacht. She was laid down in Pembroke Dockyard in November 1842 and Sir William Symonds at last had his chance to design one. She was twice the length of the *Royal George* and three times the tonnage, with an engine and paddle wheels to drive her at 11½ knots. The Queen saw her fitting out at Deptford and pronounced her 'a beautiful vessel, with splendid accommodation'. The name *Victoria and Albert* was perhaps a declaration of marital fidelity; whether consciously or not, it was a contrast to Charles I's policy of naming yachts after his mistresses. There was a precedent in the *William and Mary*, though that also proclaimed a joint monarchy. It was not unknown for yachts to be named after spouses, as in the *Royal Charlotte* and *Royal Adelaide* but *Victoria and Albert* celebrated a uniquely successful royal marriage. She would open up many new possibilities, with royal visits to Ireland and across the Channel, and to the outlying parts of Britain. Combined with the growth of the railway network and the electric telegraph, such events could be co-ordinated far more efficiently than ever before.

The *Victoria and Albert* was too large to enter some of the smaller ports, so the *Fairy* of 317 tons was built to accompany her and was launched at Blackwall in 1844. She also carried the royal family to and from their new home in Osborne House on the Isle of Wight, until the 98-ton *Elfin* of 1849 took over that duty. A new and larger *Victoria and Albert* was built at Pembroke and commissioned in 1855, more than double the size to accommodate the growing family. She would serve for the rest of the century, though less so after the death of Albert in 1861 brought the Queen into a profound and prolonged mourning. The royal yachts were now established at Portsmouth, connected by rail since 1848. Deptford Dockyard was closed in 1869.

The Prince of Wales

In August 1851 Edward, Prince of Wales, the ten-year-old eldest son of Victoria and Albert, stood on the deck of the eponymous yacht wearing a white sailor suit which was probably soiled after an excursion ashore on the Isle of Wight. They were at anchor in Alum Bay and all eyes were strained towards the Needles to the west, anxious to know who was winning a yacht race round the island. At 5.50 that afternoon the transatlantic challenger, aptly named *America*, appeared in the distance long before any of her competitors. As she passed the royal yacht she dipped her colours, though that was not expected during a race. There was a moment of doubt as the wind dropped, and it was just possible that one of the lighter competitors might overtake her, but she reached the finishing line off Cowes minutes ahead of the *Aurora*. This was the year of the Great Exhibition, perhaps the most glorious peacetime event in the history of the British Empire – yet here was Britannia being defeated on her own element by an upstart nation, amid great publicity. The winner was awarded a trophy which became known as the America's Cup and is the subject of fierce competition to this day.

The adult prince succeeded his late father as patron of the Royal Yacht Squadron at Cowes in 1863, and he began to show an interest in yachting on his own account when he purchased the 37-ton cutter yacht *Dagmar* in 1865, followed by the steam yachts *Princess* and *Zenobia*. He began racing with the purchase of the 20-ton *Hildegarde* in 1876, to be succeeded by the 104-ton *Formosa* in 1880 and the 210-ton *Aline* two years later. All of them appeared regularly at Cowes Week early in August, now established as a major event in the social calendar between the horse racing season and the start of grouse shooting in Scotland – Edward featured strongly in all of these.

The *Britannia*

At last in 1893 Edward decided to take the plunge and build his own yacht to the very highest standards. George Lennox Watson was the leading yacht designer of the age, on the eastern side of the Atlantic at least. In 1873 at the age of

The Prince of Wales, known as 'Bertie', wearing a sailor suit on board the *Victoria and Albert* in 1846. (Royal Collection Trust / © Her Majesty Queen Elizabeth II 2022, 980024.bj)

The yachts *Britannia* and *Valkyrie* racing in 1893. (Royal Collection Trust / © Her Majesty Queen Elizabeth II 2022, 450778)

twenty-two, having completed an apprenticeship in Robert Napier's shipyard in Glasgow, he set up the world's first yacht design office. He had studied hydrodynamics in some detail and attracted commissions from the Vanderbilt family, Sir Thomas Lipton the tea magnate, the Rothschilds and the German Kaiser, but his boats never succeeded in winning the America's Cup against designs by his American rival, Nathanael Herreshoff.

Watson was summoned to the royal residence at Sandringham and met the prince and his wife Alexandra. Back in Glasgow he began to draw the lines of the new yacht. He favoured a round scow-like bow instead of the conventional straight stem which had been standard up till then, and this caused a sensation, being described as 'gratuitously ugly' until the critics got used to it. She was to be 121.5ft long overall, 87.8 on the waterline and 23.3ft broad. She had the deep ballast keel which was an innovation since the last sailing royal yachts, which gave her a depth of 15.1ft – unlike the older yachts with their relatively flat bottoms, she would tilt heavily to one side if she ever went aground. She was built in D and W Henderson's yard in Glasgow, alongside Lord Dunraven's *Valkyrie*, which was similar and would be her rival in many races. She was of composite construction, with metal frame and wooden planking. She was cutter-rigged, as was normal at the time, with a single mast 142ft above the deck. The mainsail was supported by a gaff aloft and a boom below and had a total area of more than 5000sq ft. There was a topsail above that and combined with the triangular headsails she carried a total sail area of more than 10,000sq ft.

There was no obvious continuity with the older sailing yachts and her function was very different, but there were certain similarities in the layout of accommodation. The aftermost third was uninhabitable due to the shape of the hull and the cabins began with a ladies' room with its own toilet accommodation. Forward of that was the owner's cabin to port and a stateroom to starboard, with a bathroom and lockers aft. Then came the saloon, an echo of the old lords' room. A bulkhead separated this from the crew's quarters in the bows, which included the captain's room, and was centred round the galley, with eight folding cots on either side. Though it was not luxurious by royal standards, Edward would sometimes spend nights on board.

Launched by the wife of the shipyard owner on 20 April 1893 under the name of *Britannia*, she had 'a most successful spin' in the Firth of Clyde and then sailed 750 miles south and east to the Thames Estuary where she beat the *Valkyrie* in the Royal Thames Yacht Club regatta with the prince on board. She started in forty-three races in her first year, winning twenty-four and coming second in nine; but honours were roughly even against the *Valkyrie*, which won a dozen races against the prince's yacht, compared with the *Britannia*'s nine. She had glorious seasons in 1894 with thirty-six first prizes in forty-eight starts, and 1895 with thirty-eight out of fifty. There was a regular pattern with races on the French Riviera in March followed by the Thames Estuary in June. The yacht sailed round to the Mersey at the end of the month for several races in the Irish Sea, then to the Clyde. There were more races off Ireland in July, followed by Cornwall, then the highlight of Cowes in August. The prince was often seen and photographed steering her.

The prince's nephew, Kaiser Wilhelm of Germany, began to take an interest and purchased the Watson-designed *Ailsa*, which he renamed *Meteor*. In 1895, taking advantage of a misguided rule change, the Kaiser commissioned the *Meteor II* which was larger, lighter and more heavily canvassed than the *Britannia*. This began a rivalry which presaged the much more portentous competition to build dreadnought battleships, which would dominate the next decade and a half. Edward had no relish for such a competition, commenting, 'The regatta used to be a pleasant recreation for me; since the Kaiser takes command it is a vexation.' He sold the *Britannia*, which passed through several speculative hands until the prince bought it back in 1899. He competed in six races that year but won none; he sold it again to Sir William Bulkeley, who raced her once in 1899 with no success.

Victoria died at Osborne on January 1901, and her body was brought over to Southampton in the *Victoria and Albert*. She never had a chance to sail in the third *Victoria and Albert*, which was not commissioned until six months later. It was not a successful design; initially there were serious stability problems, and her Belleville boilers were soon outmoded. She did not impress the new King Edward VII, who commented: 'She may be safe but I fear she will always be ugly.' Nevertheless, she was used more than sixty times by Edward and his queen and would survive for more than half a century. The mercurial King bought the *Britannia* back in 1902 to use her for cruising up and down the Solent with the Queen and his family on board – a much quieter existence in a turbulent era. He died in May 1910.

∽ A sailor king ∾

The new King George V was the second son of Edward and as such he followed a naval career. He joined the static training ship *Britannia* (not to be confused with either royal yacht of the same name) at Dartmouth in 1877 and by 1889 he was in

Edwardian soft power seen in play with the Kaiser here depicted visiting *Britannia*. (Royal Collection Trust / © Her Majesty Queen Elizabeth II 2022, 921031)

command of a torpedo boat, followed by a gunboat the next year; but like many naval officers he had been 'round the world but never in it', and his education was seriously limited. Then in 1892 his lethargic elder brother Edward died. George, though ill and stricken with grief, married Edward's fiancée, Princess May of Teck.

George inherited the *Britannia* yacht in 1910 and used her to train his eldest sons, Edward and George, during family cruises along the Channel coast. Yacht racing had begun to revive under further rule changes, and in 1913 Major Philip Hunloke was appointed the King's Sailing Master. Though she was still fitted for cruising rather than racing, she won ten handicap races that year. In 1914 she toured the country taking part in various races, but by the time Cowes Week came round in August the United Kingdom was at war with Germany. Most of her crew were naval reservists who were called up to man cruisers. Unlike steam and motor yachts, she was not suitable to be converted for patrol duties and anti-submarine warfare, so she spent the next four years in a mud berth in the Medina River on the Isle of Wight.

By 1920 yacht racing had begun to revive again, thanks to further rule changes, and the *Britannia* was towed up to the Clyde to have her bulwarks reduced. In July the King and Queen voyaged to the Firth in the *Victoria and Albert* and the King boarded the yacht on the 12th to race her for the first time since the war. It was heavy rain or '*Britannia* weather', but His Majesty was seen to enjoy himself as she won a 39-mile race. Despite her age she was still competitive. There was near disaster off Deal in July when her boom was damaged, but it was repaired in time for Cowes Week. She won seven out of twenty-three starts in the 'big-handicap' class that year. During the winter she was re-rigged at Cowes and she raced every year between 1922 and 1935 except 1929. The year 1933 was perhaps her best, with twelve first places and twelve seconds out of thirty-nine starts. In 1931 she was re-rigged again with a modern Bermuda rig replacing the gaff, and in 1935 with a taller and narrower sail, with much higher aspect ratio, which research showed was more efficient. But there was little chance to prove it: the King died at Sandringham just before midnight on 20 January 1936. The *Britannia* was now definitely outdated as a racing yacht against the new 'J' class and suggestions of preserving her were regarded as impracticable. Much of her gear was disposed of to various institutions, and on the night of 9/10 July the destroyers *Amazon* and *Winchester* towed her down the Solent and past the Needles. She was scuttled about a mile south of St Catherine's Point on the Isle of Wight.

George V's son, Edward VIII, had little interest in yacht racing and his short reign was dominated by his affair with Mrs Wallis Simpson. During Cowes Week of 1936 he was enjoying a different kind of yachting, cruising in the eastern Mediterranean in Lady Yule's *Nahlin*, designed by the firm of G L Watson and the epitome of luxury. When he abdicated in December he was taken across the Channel into exile in a destroyer, not the royal yacht.

His brother, now King George VI, had also served as a naval officer, though it was perhaps not the ideal career in view of his nervousness and his famous stammer. He joined the naval college at Osborne in 1909 and served in the battleship *Collingwood* at Jutland in 1916, but he too had little interest in yachting. As King he visited France in the Admiralty yacht *Enchantress* and found time to go to Cowes Week with the *Victoria and Albert* in 1938. In 1939 he sailed to Canada in the liner *Empress of Australia*. The old yacht was becoming an embarrassment, but the government had doubts about the political effect of spending large sums on a new one during the Great Depression. Outline plans for a replacement were prepared, but progress was interrupted by the start of war in September 1939. Again, yachts were taken up for naval duties, but not the *Victoria and Albert*; and hundreds of yachtsmen were given hasty training to become naval officers.

Warships rather than yachts were now used for long overseas visits. The future kings Edward VIII and George VI had sailed around the empire in the battlecruiser *Renown* in 1927. And George, now King, sailed to South Africa in the new battleship *Vanguard* in 1947.

A new yacht was already in contemplation, though there was still the problem of avoiding conspicuous consumption in the post-war age of austerity. One of the first tasks of the new Queen Elizabeth II after the sudden death of her father in February 1952 was to discuss a yacht to be built by John Brown's on the Clyde. Though the country had entered a new and more ebullient age, the press announcement in November was cautious. 'Her Majesty the Queen has seen and approved a model showing the general arrangement of the small hospital ship to be used as a Royal Yacht in peacetime.' Launched in April 1953, less than two months before the Coronation, she was named *Britannia*, which symbolised the nation rather than the monarchy as previous yachts had done. She was 412ft long and displaced 4961 tons, no larger than the latest *Victoria and Albert*, but much better designed. Her turbine engines could drive her at a speed of 22.5 knots. But she was not ready in time for the Coronation fleet review at Spithead on 15 June. The royal family toured the ships in the converted frigate *Surprise*.

Commissioned early in 1954, the *Britannia* literally added a new dimension to royal yachting, for she was capable of ocean voyages. She carried the Duke of Edinburgh on a world tour in 1956, taking in the Olympic Games in Melbourne. She was the first ship to sail up the St Lawrence Seaway in 1959. In the new age of air travel she could sail to a site, to be joined by the royals later. In all she made over six hundred overseas visits, eighty-five state visits and 110 UK visits. Less happily, she was used for the honeymoons of Princess Margaret and Anthony Armstrong-Jones, Princess Anne and Captain Mark Phillips, Prince Charles and Lady Diana Spencer, and Prince Andrew and Sarah Ferguson – all of which ended in divorce. Her last task was to wind up the British presence in Hong Kong in 1997, after which she was, controversially, withdrawn from royal service and laid up as a museum ship in Leith, not far from where George IV and Victoria had landed in 1822 and 1842.

The Duke of Edinburgh

Prince Philip, Duke of Edinburgh, had a truncated naval career like William IV, George V and George VI, but he did not have the satisfaction of becoming king afterwards – instead he devoted his life to supporting his wife Elizabeth after her sudden accession, and to various good causes including many maritime ones, for he never lost his interest in sea affairs. When the couple were married in 1948, the Island Sailing Club at Cowes presented them with the *Bluebottle*, a Dragon-class boat of a type which was designed by Johan Anker in Sweden in 1929 and had just gained Olympic status. The duke's second boat was the Flying Fifteen *Coweslip*, designed by the ebullient, adventurous Uffa Fox, who had also designed the airborne lifeboat to be dropped from an aircraft to rescue downed aviators in the Second World War, and many types of power and sailing boat. He pioneered the idea of a planing dinghy, with the crew leaning over the side in a trapeze to prevent her capsizing. *Coweslip*, clearly named after her port of origin, was presented by the town to the royal couple; her designer became a firm friend of the duke, and often crewed for him. She was 15ft long on the waterline and 20ft overall. She was shipped out to Malta in the aircraft carrier *Glori*, where the prince was first lieutenant of the destroyer *Chequers*; he was joined there by Princess Elizabeth.

The boat and the duke were back home in time for Cowes Week in 1951, when they started as usual on a signal from one of the old *Royal Adelaide*'s guns and won by seventeen seconds. But, according Fox, Cowes Week was but a shadow of its former self:

> The price we have had to pay for enduring and winning two dreary wars is reflected in the size of craft racing. Before the 1939 war the smallest international class was the 6-metre and this year the largest international classes are ten 6-metres, and there are only some half dozen of these. Broadly speaking, the position with yachting today is that there are a greater number of yachtsmen and yachts than ever before, but all of them in smaller craft.

In these circumstances, it was acceptable to race in boats with crews of two each, 29ft and 20ft long respectively. His grandfather-in-law had raced in a 121ft yacht with a crew of more than thirty.

Fox described the duke in the 1952 race, his first time on board since his wife became queen:

> There are a great many good helmsmen, but few of those are able to tend the sheets and gear as Prince Philip did this day. Throughout this run he had the tiller, mainsheet and spinnaker guy, steering and handling his sheets to perfection. We were racing against first-rate helmsmen and crew who had been sailing together all summer and who knew their boats and gear as well as the back of their hand.

They finished second in the race. Then the duke raced *Bluebottle*. After a well-judged start in very light winds, '[b]y bearing away to leeward and sailing quietly away to leeward, he soon pulled out into first place and led almost all the way across the Solent to the first mark and again elected to round the buoy to leeward of two other boats'. But on a subsequent leg, '[w]ith our sheets slightly eased we ramped across the Solent, as by now the wind had increased, rounding Old Castle Point third, and, setting the spinnaker, we finished third.'

In 1953, Coronation year, *Bluebottle* was put out of the race by a broken stay. In 1955 *Coweslip* finished fourth. In 1957 the duke was joined by the eight-year-old Prince Charles, who was reported to have enjoyed himself, though he looked very glum in newsreels. When the duke was unable to attend in 1959 the Cowes tradespeople lamented the lack of business. A teashop owner said, 'Without the Duke, I expect my trade to drop. Most of us in this line of business rely for our profits on the thousands of day trippers who flock to Cowes in the hope of seeing royalty.' An inn manageress commented, 'The yachting people will still come in force, but it is the visitors and day trippers we want.' The duke was back by 1960 but in 1962 *Coweslip* capsized and her crew had to be rescued. Undaunted, the duke raced *Coweslip* three times in 1963.

By that time he had loaned *Bluebottle* to the Royal Naval College at Dartmouth and purchased a very different yacht. *Bloodhound* was 63ft long, more than three times the size of the boat she replaced. She was designed by Charles E Nicolson, of the famous Camper and Nicolson yard in 1939, for ocean racing. The duke and Fox raced her at Cowes in 1962; in 1964 Philip took his daughter Princess Anne out in *Bloodhound*, but she would come to prefer horses, competing in the 1972 Olympics. But royal interest in racing was declining and the *Bloodhound* would be used mainly for cruising, accompanying the *Britannia* on the annual visit to the Western Isles of Scotland each summer. She was sold in 1969 and fell into disrepair before being restored and set beside the *Britannia* in Leith in 2010.

There are occasional proposals to build a new royal yacht as a showpiece for British enterprise, though the world has changed much since the 1950s. The empire has gone and travel is mainly by air. Super yachts are the province of billionaires, and the largest racing yachts are mostly sponsored by corporations. Even with the £280 million proposed for the latest project, it would be difficult to compete with the largest private yachts. Is the age of the royal yacht finally over?

Prince Phillip and Uffa Fox sailing the Flying Fifteen *Coweslip*.

SECOND WIND

Bibliography

General history
Acts of Parliament, *New Churches in London and Westminster*, 9 Anne cap 17
Burnet, Gilbert, *History of his Own Time*, vol II, Oxford 1823
Campbell, R, *The London Tradesman*, London, 1747
Crane, John, *Cursory Observations on Sea-Bathing*, Weymouth, nd
Friel, Ian, *Henry V's Navy*, 2015, London
Giles, J A (ed), *Chronicles of the Crusades*, London, 1848
Hayton and Cruikshanks (eds), *The History of Parliament, The House of Commons 1690–1715*
Hutton, Ronald, *The Restoration*, Oxford, 1987
Morillo, Stephen, *The Battle of Hastings*, 1998
Ogg, David, *England in the Reign of Charles II*, 2 vols, Oxford, 1934
———, *England in the Reigns of James II and William III*, Oxford, 1969
The Royal Kalendar, London, 1797
Watson, J Steven, *The Reign of George III, 1760–1815*, Oxford, 1960
Trevelyan, G M, *England under Queen Anne*, 3 vols, London, 1931/2
Williams, Basil *The Whig Supremacy*, Oxford, 1962

Royal history and biography
Aspinall, A (ed), *The Correspondence of George, Prince of Wales*
———, *The Later Correspondence of George III*, vol I, Cambridge, 1962
Buckingham and Chandos, Duke of, *Memoirs of the Court of George IV, 1820–1830*, London, 1859
Chedzoy, Alan, *Seaside Sovereign, King George III at Weymouth*, Wimborne, 2003
Hatton, Ragnhild, *George I*, London, 1978
Hedley, Owen, *Queen Charlotte*, London, 1975
Lower, Sir William, *Relation … of the Voyage and Residence … which … Charles II hath Made in Holland …*, The Hague, 1667
Pepys, Samuel, *His Majesty Preserved …*, reprinted London, 1956
Thompson, Andrew C, *George II, King and Elector*, Yale, 2012

Naval and shipbuilding history
Anon, *Further Advice to a Painter, or Directions to Draw the Late Engagement, Aug 11th, 1673*
Burney, W, *Universal Dictionary of the Marine*, London, 1815
Chapelle, Howard, *The Search for Speed under Sail*, New York, 1967
Chapman, Frederick af, *Architectura Navalis Mercatoria*, reprinted 2006
Cornford, L Cope, *A Century of Sea Trading*, London, 1924
Coronelli, Vincenzo, *Navi o Vascelli (Ships and Vessels)*, Venice, 1697, copy in the National Maritime Museum Caird Library De Groot and Vorstman, *Sailing Ships*, Amsterdam, 1980
Derrick, Charles, *Memoirs of the Rise and Progress of the Royal Navy*, London, 1806
Fox, Frank, *Great Ships*, Greenwich, 1908
Ingram, Bruce S (ed), *Three Sea Journals of Stuart Times*, London, 1936
Lavery, B (ed), *Deane's Doctrine of Naval Architecture*, London, 1981
———, *The Ship of the Line*, vol I, London, 1983
———, *The Arming and Fitting of English Ships of War*, London, 1987
McLaughlin, Ian, *Sloop of War*, Barnsley, 2014
Powley, Edward B, *The English Navy in the Revolution of 1688*, London, 1928
Winfield, Rif, *British Warships in the Age of Sail, 1714–1792*, Barnsley, 2014

Topography
Anon, *England's Gazetteer*, vol II, London, nd, np
Norie, J W, *Sailing Directions for the Navigation of the North Sea*, London, 1846
Nugent, Thomas, *The Grand Tour, or, A journey through the Netherlands, Germany, Italy and France*, London, 1756

Royal yachts
Dalton, Tony, *British Royal Yachts*, Tiverton 2002
Gavin, Charles, *Royal Yachts*, London, 1932
Grigsby, J E, *Annals of our Royal Yachts, 1604–1903*, London, 1953
Irving, John, *The King's Britannia*, London, c1936
McGowan, Alan, *Royal Yachts*, Greenwich,
Tanner, Matthew, *Royal Yacht Mary*, Liverpool, 2009

Other biography
Charnock, John, *Biographia Navalis*, vol 1, London, 1794
Churchill, Winston, *Marlborough, His Life and Times*, Book I, London, 1947
Grierson, H (ed), *The Letters of Sir Walter Scott*, vol VII, 1934
Knighton, Lady (ed), *Memoirs of Sir William Knighton*, vol I, Paris, 1838
McCann, Timothy J (ed), *The Correspondence of the Dukes of Richmond and Newcastle, 1724–1750*, Sussex Record Society, Gloucester, 1984
Murray, Sir G (ed), *The Letters and Dispatches of John Churchill, First Duke of Marlborough*, 1845
O'Byrne, William R, *A Naval Biographical Dictionary*, London, 1849
Oxford Dictionary of National Biography
Rodger, N A M, *The Insatiable Earl*, London, 1993
Ross, Sir John (ed), *Memoirs of Admiral Lord de Saumarez*, vol I, 1838
Tomalin, Claire, *Samuel Pepys, The Unequalled Self*, London, 2001
Walpole, Horace, *The Works of Horace Walpole*, vol VI, London, 1822
Wellington, Arthur Wellesley, Duke of, *Despatches, Correspondence and Memoranda of Field Marshal Arthur Duke of Wellington*, London, vol IV, 1867–80

Art
National Maritime Museum, *Concise Catalogue of the Oil Paintings in the National Maritime Museum*, Woodbridge, 1988
Robinson, M S, *Van de Velde Drawings; a Catalogue of the Drawings in the National Maritime Museum*, Cambridge, 1973
———, *The Paintings of the Willem Van de Veldes*, vol II, Greenwich, 1990
Weber, Richard (ed), *The Willem Van de Velde Drawings in the Boymans-van Beuningen Museum, Rotterdam*, ed 3 vols, 1979

Articles
Endsor, Richard, 'The Van de Velde Paintings for the Royal Yacht *Charlotte*', *Mariner's Mirror*, vol 94, pp271–5
Herman, Neil, 'Henry Grattan, the Regency Crisis and the Emergence of a Whig Party in Ireland, 1788–9', *Irish Historical Studies*, vol 32, no 128 (Nov 2001)
Lavery, B, 'The Rebuilding of British Warships', *Mariner's Mirror*, vol 66, pp5–16, 113–26
Moneypenny, Effie, and David Antscherl, 'A Restoration Yacht's Design Secrets Unveiled', *Mariner's Mirror*, vol 107, pp164–86
Moneypenny, Effie, and Simon Stephens, 'A Model of the Royal Yacht *Henrietta*', *Mariner's Mirror*, vol 104, pp172–91
Moneypenny, Kelvin, 'The Royal Yacht *Henrietta* of 1679', *Mariner's Mirror*, vol 100, pp132–46
Ryan, W F, 'Peter the Great's English Yacht', in *Mariner's Mirror*, vol 69, pp65–87

Navy Records Society publications
Anderson, R C (ed), *Journal of the Earl of Sandwich*, 1928
———, *The Journals of Sir Thomas Allin*, vol I, 1939
———, *Journals and Narratives of the Third Dutch War*, 1946
Bonner-Smith, D, *The Barrington Papers*, vol I, 1937
———, *Captain Boteler's Recollections*, 1942
Browning, Oscar, *The Journal of Sir George Rooke, 1700–1702*, 1897
Hattendorf et al (eds), *British Naval Documents, 1204–1960*, 1993
Laughton, J K (ed), *The Naval Miscellany*, vol II, 1912
Powell, J R, and E K Timings (eds), *Documents Relating to the Civil War*, 1963
———, *The Rupert and Monck Letter Book 1666*, 1969
Tanner, J R (ed) *Catalogue of the Pepysian Manuscripts*, 4 vol 1, 1903–23,
———, *Samuel Pepys' Naval Minutes*, 1926

State papers
Calendar of State Papers, Domestic, Charles II, ed M A E Green, F H B Daniell, F Bickley, 28 vols, London, 1860–1947
Calendar of State Papers, Domestic, Charles II Calendar of Treasury Books, vol 7, 1681–1685, London, 1916
Calendar of State Papers, Domestic, James II, ed E K Timings, 3 vols, London, 1960–72
Calendar of State Papers, Domestic, William and Mary, ed W J Hardy, E Bateson, 11 vols, London, 1895–1937

Other printed original sources
A Collection of the State Papers of John Thurloe, London, 1742
The Diaries of Samuel Pepys, ed Robert Latham and William Matthews, 11 vols, London, 1970–83
Historical Manuscripts Commission, *The Manuscripts of the Earl of Dartmouth*, vol III, 1896
Historical Manuscripts Commission, *Report on the Manuscripts of Lady Du Cane*, London, 1905

Manuscripts
The National Archives
ADM 1 In Letters, ADM 2 Orders and Instructions, ADM 3 Admiralty Minutes, ADM 36, musters, ADM 51, logs, ADM 106, Navy Board Papers. Many of the in-letters have been catalogued and summarised, of great benefit to historians. The series also includes the Deptford Yard Letter Books which are uncatalogued but contain many gems.
The PROB series includes the wills of some of the yachts' officers and crews.
State Paper series, including SP 36.

National Maritime Museum
Lieutenant's Logs, ADM/ L series
ADM/A series
ADM/B series
SAN, Earl of Sandwich papers
Ships Plans in the Brass foundry at Woolwich, mostly in the NPD series.

Royal Collection
GEO/ADD/43/3e, Diaries of Queen Charlotte
GEO/ADD/2/93a, Journal of John Hughes on the voyage of Princess Charlotte, 1761
GEO/ADD/45/13-13a, Accounts of table expenses aboard the Royal Sovereign

Other archives
Lincolnshire Archives, Journals of Christopher Gunman, JARVIS/9/1/A/5

Notes

PART I SAILOR KINGS
1 The Origin of the Yacht
1. *Victoria County History, Berkshire*, vol 3, London, 1923, p413.
2. William of Poitiers quoted in Stephen Morillo, *The Battle of Hastings*, 1998, p9.
3. Chronicle of Richard of Devizes, in *Chronicles of the Crusades*, ed Giles, London, 1848, p13.
4. Ian Mortimer, *1415: Henry V's Year of Glory*, London, 2009, p338.
5. Ian Friel, *Henry V's Navy*, 2015, London, pp10–11.
6. De Groot and Vorstman, *Sailing Ships*, Amsterdam, 1980, p56.
7. Navy Records Society, *Documents Relating to the Civil War*, ed Powell and Timings, 1963, p380.
8. Sir William Lower, *Relation ... of the Voyage and Residence ... which ... Charles II hath Made in Holland ...*, The Hague, 1667, p26.
9. Samuel Pepys, *His Majesty Preserved ...*, reprinted London, 1956, p42.
10. Lower, op cit, pp27–8.

2 The First English Yachts
1. National Maritime Museum, PAH1887, PAF6945, BHC3600.
2. Navy Records Society, *Catalogue of the Pepysian Manuscripts*, vol 1, 1903, ed J R Tanner, pp294–5.
3. Lower, op cit, p26.
4. Ibid, pp48–9.
5. Quoted in Matthew Tanner, *Royal Yacht Mary*, Liverpool, 2009.
6. British Museum Prints, 1874, 0808.97.
7. Pepys, *Diary*, 8 November 1660.
8. Alan McGowan, *Royal Yachts*, Greenwich, 1977, p2.
9. Thurloe State Papers, vol 5, pp16–17.
10. Navy Records Society, *Samuel Pepys' Naval Minutes*, ed J R Tanner, 1926, p9.
11. Pepys, *Diary*, 12 January 1661.
12. *Deane's Doctrine of Naval Architecture*, ed Lavery, 1981, p112.
13. National Archives, ADM 106/324/405.
14. Calendar of State Papers, 1664–5, pp326–40.
15. Pepys, *Diary*, 13 8 62.
16. Ibid, 5 September 1662.
17. 106/323/149, 106/324/405.
18. Diary, 2 March 1663.
19. National Archives, ADM 106/285/88.
20. Navy Records Society, *Catalogue of the Pepysian Manuscripts*, vol 3, 1909, ed J R Tanner, p44.

3 The Uses of the Yachts
1. Navy Records Society, *Catalogue of the Pepysian Manuscripts*, vol 4, 1923, ed J R Tanner, p258.
2. Pepys, *Diary*, 1 October 1661.
3. Navy Records Society, *Samuel Pepys' Naval Minutes*, ed J R Tanner, 1926, p116.
4. Navy Records Society, *Journal of the Earl of Sandwich*, ed R C Anderson, 1928, p82.
5. Ibid, pp138–9.
6. *Calendar of State Papers, Domestic*, 1677–8, London, 1911, 6 November 1677.
7. Ibid, 12 November 1677.
8. Navy Records Society, *The Journals of Sir Thomas Allin*, ed R C Anderson, vol 1, 1939, p71.
9. *Journal of the Earl of Sandwich*, op cit, p89.
10. *The Journals of Sir Thomas Allin*, vol 2, p30.
11. Ibid, vol 1, pp96–7.
12. *Calendar of State Papers, Domestic*, 1668–9, London, 1894, October 1668.
13. Ibid, 1671, London, 1895, 21 July.
14. B Lavery, *The Ship of the Line*, vol 1, London, 1983, pp42–3.
15. *Calendar of State Papers, Domestic*, 1663–4, London, 1862, 19 February 1664.
16. *Catalogue of the Pepysian Manuscripts*, vol 4, p581.
17. National Maritime Museum Manuscripts, OBK/8.
18. *Catalogue of the Pepysian Manuscripts*, vol 2, op cit, p211.
19. Ibid, vol 3, p263.
20. *Calendar of Treasury Books*, vol 7, 1681–1685, London, 1916, 14 February 1682.
21. Ibid, vol 4, 1672–75, London, 1909, 4 October 1675.
22. Ibid, vol 7, 1681–5, appendix, June 1683.

4 The Later Yachts
1. *Calendar of State Papers, Domestic, 1663*, London, 1862, 6 September 1633.
2. *The Willem Van de Velde Drawings in the Boymans-van Beuningen Museum, Rotterdam*, ed Richard Weber, 3 vols, 1979, vol I, pp63, 66.
3. *Calendar of State Papers, Domestic, 1670*, London, 1895, 8 August 1670.
4. National Maritime Museum Manuscripts, OBK/8.
5. *Calendar of State Papers, Domestic, 1670*, 14 April 1670.
6. Tony Dalton, *British Royal Yachts*, Tiverton, 2002, p49.
7. *Catalogue of the Pepysian Manuscripts*, vol 2, p9.
8. Ibid, p313.
9. Ibid, vol 4, pp246–7.
10. Ibid, p228.
11. National Archives, ADM 106/310.
12. Ibid, ADM 106/307.
13. B Lavery, *The Ship of the Line*, vol 1, London, 1983, pp42–3.
14. *Samuel Pepys' Naval Minutes*, pp116, 128.
15. *Catalogue of the Pepysian Manuscripts*, vol 4, pp391, 408.
16. Ibid, p462.
17. Ibid, p585.
18. Ibid, pp638–9.
19. National Archives, ADM 106/346 passim.
20. Ibid, 106/350/168, 171.
21. Dalton, from ADM 106/362.
22. Narcissus Lutterell, *A Brief Historical Relation of State Affairs*, Oxford, 1857, vol 1, p214.
23. Dalton, ADM 106/362.
24. National Archives, ADM 106/360.
25. Lutterell, op cit, vol 1, p218.
26. Navy Records Society, *The Tangier Papers of Samuel Pepys*, ed Edwin Chapell, 1935, p305.
27. Navy Records Society, *British Naval Documents, 1204–1960*, ed Hattendorf et al, 1993, p267.

5 The Loss of the *Gloucester*
1. Based on the Journals of Christopher Gunman in the Lincolnshire Archives, JARVIS/9/1/A/5. John Charnock, *Biographia Navalis*, vol 1, London, 1794, p228n.
2. *Samuel Pepys' Naval Minutes*, op cit, pp149–50.
3. Ibid, pp149–50.
4. Charnock, op cit, p152n.
5. Quoted in Winston Churchill, *Marlborough, His Life and Times*, Book One, London, 1947, p158.
6. Gilbert Burnett, *History of his Own Time*, vol II, Oxford 1823, p315.
7. *Calendar of State Papers, Domestic*, 1682, 13 May 1682.
8. Ibid, June 1682.
9. *Samuel Pepys' Naval Minutes*, op cit, p147.
10. National Archives, PROB 11/379/471.

6 The Yachts of War
1. Navy Records Society, *The Rupert and Monck Letter Book 1666*, ed Powell and Timings, 1969, pp81, 88.
2. Ibid, p156.
3. *The Journals of Sir Thomas Allin*, vol 1, op cit, p223.
4. Op cit, pp264–5.
5. NRS *Journal of the Earl of Sandwich*, op cit, p293.
6. Navy Records Society, *Journals and Narratives of the Third Dutch War*, ed R C Anderson, 1946, pp4–5.
7. Ibid, pp69–70.
8. Ibid, pp78, 79.
9. Ibid, pp108–9.
10. Ibid, pp151–3.
11. Ibid, p315.
12. Ibid, pp317–18.
13. Ibid, p305.
14. *The Willem Van de Velde Drawings in the Boymans-van Beuningen Museum*, op cit, vol ii, pp61–5.
15. *Journals and Narratives of the Third Dutch War*, op cit, pp324–6.
16. *The Willem Van de Velde Drawings in the Boymans-van Beuningen Museum*, op cit, vol ii, pp65–8.
17. *Journals and Narratives of the Third Dutch War*, op cit, p326.
18. Anon, *Further Advice to a Painter, or Directions to Draw the Late Engagement, Aug 11th, 1673*, p5.
19. National Archives, 106/384/429.
20. Edward B Powley, *The English Navy in the Revolution of 1688*, London, 1928, pp28–9.
21. Historical Manuscripts Commission, *The Manuscripts of the Earl of Dartmouth*, vol III, 1896, p61.
22. Ibid, p67.
23. Ibid, p138.
24. Powley, op cit, p137n.
25. M S Robinson, *The Paintings of the Willem Van de Veldes*, vol 2, Greenwich, 1990, 1000-3.
26. Narcissus Lutterell, op cit, vol 1, p501.

PART II TO THE WARS
1 The Changing Role
1. National Maritime Museum Manuscripts, ADM/L/C/281.
2. E B Powley, op cit, pp356–7.
3. *The Life and Glorious Actions of the Right Honourable Sir George Rook*, London, 1707, pp11–12.
4. *Oxford Dictionary of National Biography*, Sir George Rooke.
5. Lutterell, op cit, vol 2, pp154–6, 159, 162–3, 165.
6. *Calendar of State Papers, Domestic*, 1690–91, London, 1808, 7 February 1691.
7. Ibid, 6 October.
8. Lutterell, op cit, vol 2, p296.
9. Ibid, p591.
10. National Maritime Museum Manuscripts, ADM A/1841/136.
11. The *History of Parliament, The House of Commons 1690–1715*, ed Hayton and Cruikshanks, entry for Cloudesley Shovell.
12. *Oxford Dictionary of National Biography*, Sir David Mitchell.
13. Luttrell, op cit, vol 3, p470.
14. Ibid, p215.
15. *Calendar of State Papers, Domestic*.
16. Lutterell, op cit, vol 2, p600.
17. Luttrell, op cit, vol 4, p288.
18. National Archives, ADM 51/585.
19. *The Life of William Fuller*, London, 1701.
20. Lutterell, op cit, vol 4, p273.
21. W F Ryan, 'Peter the Great's English Yacht', in *Mariner's Mirror*, vol 69, Feb 1983, pp65–84.
22. National Archives, ADM 51/585.
23. Ibid.
24. Ibid.
25. *Calendar of State Papers, Domestic*, 1700–02, London, 1937, 11 July 1700.
26. From Navy Records Society, *The Journal of Sir George Rooke*, ed Oscar Browning, 1897, passim; *Mary* log, National Archives, ADM 51/585.

2 New Yachts
1. *Catalogue of the Pepysian Manuscripts*, op cit, vol 1, p77.
2. National Maritime Museum Manuscripts, ADM/A/1803-4.
3. Ibid, ADM/A/1810.
4. Ibid.
5. Ibid, ADM/A/1811.
6. National Archives, ADM 106/448/287, 447/316.
7. Ibid, ADM 106/448/287.

8. National Maritime Museum Manuscripts, ADM/A/1821, 106/560.
9. National Archives, ADM 106/494/201, 484/106, 494/324.
10. Charles Derrick, *Memoirs of the Rise and Progress of the Royal Navy*, London, 1806, p157.
11. National Maritime Museum Manuscripts, ADM/A/1828/134.
12. Ibid, ADM/A/1829/1.
13. Ryan, *Peter the Great's English Yacht*, op cit, p70.
14. National Maritime Museum Manuscripts, ADM/A/1834/14.
15. Ryan, op cit, p68.
16. Ibid, p66
17. Ibid, pp67–8.
18. National Maritime Museum Manuscripts, ADM/A/1841/161.
19. Ryan, op cit, p66.
20. Ibid, p75.
21. Bruce S Ingram, ed, *Three Sea Journals of Stuart Times*, London, 1936, pp116–17.
22. National Archives, ADM 106/526/339.
23. Howard Chapelle, *The Search for Speed under Sail*, New York, 1967, pp36–46.
24. National Archives, ADM 106/560/170.
25. Ian McLaughlin, *Sloop of War*, Barnsley, 2014, pp97–9.
26. National Archives, ADM 106/561/254.
27. Ibid, ADM 1/3595.
28. *Calendar of State Papers, Domestic*, 1703–4, London, 1924, 30 June 1703.
29. National Archives, ADM 106/3296.
30. Ibid, ADM 106/3996.

3 Marlborough at Sea
1. *London Gazette*, 11–15 April 1706, 18–21 April 1709, 6–9 June 1709.
2. Navy Records Society, *The Journal of Sir George Rooke, 1700–1702*, ed Oscar Browning, 1897, pp150–239.
3. *The Letters and Dispatches of John Churchill, First Duke of Marlborough*, ed Sir G Murray, 1845, vol 1, pp54–5n.
4. Ibid, p563.
5. National Archives, ADM 51/585.
6. Dispatches, op cit, p249.
7. Ibid.
8. Ibid, pp 337–45; National Maritime Museum Manuscripts, L/C/286.
9. *Dispatches*, op cit, p698.

PART III THE HANOVER CONNECTION
1 The Coming of King George
1. Basil Williams, *The Whig Supremacy*, Oxford, 1962, p50.
2. Quoted in *Oxford Dictionary of National Biography*.
3. National Archives, ADM/13
4. National Archives, ADM 51/4143.
5. *Oxford Dictionary of National Biography*.
6. Compiled from National Archives, ADM 8/14, 15, 16.
7. *Caledonian Mercury*, 20 June 1723.
8. *The Newcastle Courant*, 5 June 1725.
9. Thomas Nugent, *The Grand Tour, or, A journey through the Netherlands, Germany, Italy and France*, London, 1756, p148.
10. Historical Manuscripts Commission, *Report on the Manuscripts of Lady Du Cane*, London, 1905, p18.
11. National Archives, ADM 51/4138.
12. Ibid, ADM 1/4196.
13. Anon, *England's Gazetteer*, vol 2, London, nd, np, under 'Margate'.
14. Ragnhild Hatton, *George I*, London, 1978, pp280–3.

2 The Yachts at Deptford and Greenwich
1. Diary, 1 July 1662, 5 August 1662, 8 August 1662, 25 August 1662.
2. National Archives, ADM 106/3317.
3. Ibid, 1/2649.
4. Navy Records Society, *Catalogue of the Pepysian Manuscripts*, op cit, vol 4, p382.
5. National Archives, ADM 106/3466.
6. Ibid, 51/586.
7. Ibid, 106/3303.
8. National Maritime Museum manuscripts, ADM/B/180.
9. Ibid.
10. National Archives, ADM 106/1198/467.
11. Navy Records Society, *The Naval Miscellany*, vol II, ed J K Laughton, 1912, p146.
12. Ibid, p147.
13. National Archives, ADM 51/586.
14. *Naval Miscellany*, op cit, vol II, p147.
15. Rif Winfield, *British Warships in the Age of Sail, 1714–1792*, Barnsley, 2014, p361.
16. National Archives, ADM 106/3308.
17. Ibid, 106/3303.
18. B Lavery, *The Arming and Fitting of English Ships of War*, London, 1987, pp57–8.
19. National Archives, ADM 180/13.
20. Burney, *Universal Dictionary of the Marine*, London, 1815, p453.
21. *British Warships in the Age of Sail*, op cit, p356.
22. National Archives, ADM 354/178, 106/840.
23. Winfield, op cit, pp101–2; National Archives, ADM 51/586.
24. Ibid, 51/586.
25. Quoted in Neal, p88.
26. National Archives, ADM 106/3300.
27. Neal, p71.
28. National Archives, ADM 106/3466.
29. Daniel Defoe, *A Tour Through the Whole Island of Great Britain*, ed P N Furbank et al, New Haven, 1991, p43.
30. Ibid.

3 Rebuild, Repair, Renovation and Replacement
1. B Lavery, 'The Rebuilding of British Warships', *Mariner's Mirror*, vol 66, pp5–16, 113–26.
2. National Archives, ADM 1/3637.
3. Ibid, 3/35.
4. Ibid, 106/3301.
5. Ibid, 106/3308.
6. Ibid, 51/4316.
7. National Maritime Museum manuscripts, SAN/V/1.
8. Ibid, 2/73.
9. National Maritime Museum, Ships Plans, APD 0962.
10. National Archives, ADM 106/3316, 2585

4 Other Duties
1. National Archives, ADM 51/586.
2. National Maritime Museum manuscripts, ADM/B/123/138.
3. Letters 1 172–3.
4. National Maritime Museum manuscripts, ADM B/124/189.
5. National Archives, ADM 2/72.
6. Ibid, 2/72.
7. R Campbell, London, 1747, p197.
8. National Archives, ADM 2/73.
9. National Maritime Museum manuscripts, ADM/B/142/220.
10. National Archives, ADM 8/28.
11. Ibid, 51/586.
12. Ibid, 51/586.
13. Ibid, 51/586.
14. Ibid, 51/586.
15. Ibid, 36/1209.
16. Ibid, 51/586.
17. Ibid, 2/77.
18. Ibid, 2/71.
19. Ibid.
20. National Maritime Museum manuscripts, ADM/B/144.
21. Ibid, ADM/B/189.
22. National Archives, ADM 106/3381.
23. Ibid, 2/71.

5 The Last King in Battle
1. Andrew C Thompson, *George II, King and Elector*, Yale, 2012, pp86–7.
2. National Archives, SP 36/15/31.
3. Ibid, ADM 51/203.
4. *The Works of Horace Walpole*, vol V, London, 1822, pp497–9.
5. John Lord Hervey, *Memoirs of the Reign of George the Second*, vol II, London, 1848, p222.
6. Ibid, pp565ff.
7. National Archives, ADM 51/3848.
8. Thompson, *George II*, op cit, p148.
9. National Maritime Museum manuscripts, Log L/R/274.
10. Sussex Record Society, *The Correspondence of the Dukes of Richmond and Newcastle, 1724–1750*, ed Timothy J McCann, Gloucester, 1984, pp170–1.
11. National Archives, ADM 1/87.
12. Ibid, 2/71.
13. Ibid, 1/579.
14. Ibid, 2/72.
15. Ibid, 1/579.
16. Ibid, 51/4316.
17. Ibid, 51/4316.
18. Ibid, 2/73.
19. Navy Records Society, *The Barrington Papers*, ed D Bonner-Smith, vol 1, 1937, pp78–85.
20. National Archives, ADM 1/579.
21. *The Works of Horace Walpole*, vol VI, London, 1822, p383.
22. Thompson, *George II*, op cit, p239.

PART IV MARRIAGE, BUSINESS AND LEISURE
1 A New Queen
1. *Newcastle Courier*, 18 July 1761.
2. *Bath Chronicle*, 10 September 1761.
3. Olwen Hedley, *Queen Charlotte*, London, 1975, p18; National Archives, ADM 106/3317.
4. Hedley, op cit, p32.
5. Historical Manuscripts Commission, Reports on Manuscripts in Various Collections, 1914, p178.
6. Hedley, *Queen Charlotte*, p37.
7. Ibid, p39.
8. National Archives, ADM 1/519.

2 Sandwich's Visitations
1. National Archives, ADM 7/658.
2. Ibid, 106/1204/193.
3. Ibid, 106/1194/318.
4. Ibid, 106/1203/238.
5. Ibid, 7/659.
6. Ibid, ADM 106/1207/206.
7. National Maritime Museum manuscripts, SAN/V/6.
8. Ibid.
9. British Library manuscripts, Kings 43.
10. National Maritime Museum manuscripts, SAN/V/7, SAN/45/C, SAN/F/4; *London Gazette*, 26 June 1773.
11. Ibid, SAN/V/7.
12. Ibid, SAN/V/8; National Archives, ADM 52/1163.
13. National Maritime Museum manuscripts, ADM L/P/306.
14. Ibid, SAN/V/11.

3 An Unhappy Marriage
1. *The Royal Kalendar*, London, 1797, p96.
2. National Archives, ADM 36/12670.
3. Ibid, 52/3220.
4. Ibid.

4. The Yachts at Weymouth
1. John Crane, *Cursory Observations on Sea-Bathing*, Weymouth, nd, p87.
2. *The Correspondence of George, Prince of Wales*, ed Aspinall, vol 3, p457.
3. *The Later Correspondence of George III*, ed Aspinall, vol 1, Cambridge, 1962, p433.
4. *Memoirs of Admiral Lord de Saumarez*, ed Sir John Ross, vol I, 1838, pp144–5.

5. Royal Archives, GEO/ADD/43/3e.
6. Ibid.
7. *The Correspondence of George, Prince of Wales*, op cit, pp358, 369.
8. Ibid, p457.
9. Ibid, p477.
10. *The Later Correspondence of George III*, vol 3, p131.
11. Ibid, p146.
12. Ibid, p553.
13. Ibid, p583.
14. Hedley, *Queen Charlotte*, op cit, p212.
15. National Maritime Museum manuscripts, ADM/L/R/303.
16. National Archives, ADM 36/16596.
17. *The Later Correspondence of George III*, vol 4, p351

5 New Yachts for a New Century
1. National Archives, ADM 106/3326.
2. Ibid, ADM 106/3472.
3. Ibid, ADM 106/3326.
4. Ibid.
5. Ibid – misdated as 1801.
6. Ibid, ADM 106/3328.
7. Ibid, ADM 106/3329.
8. Ibid, ADM 180/13.
9. Charles Gavin, *Royal Yachts*, London, 1932, p96.
10. Navy Records Society, *Captain Boteler's Recollections*, ed D Bonner-Smith, 1942, pp93, 97, 100–1.
11. National Archives, ADM 180/13.
12. British Newspaper Archive, passim, 1819.

6 The End of the Wars
1. *Sun* (London), 25 April 1814.
2. *Public Ledger and Daily Advertiser*, 28 April 1814.
3. Royal Collection, GEO/MAIN/45076.
4. Ibid, GEO/MAIN/45116-45117.
5. Ibid, GEO/MAIN/45124-45125.
6. *Brighton Herald*, 13 September.
7. *Kentish Weekly Post or Canterbury Gazette*, 16 September 1817.
8. *Oxford University and City Herald*, 20 September 1817.
9. *Star* (London), 14 August 1819.
10. *Exeter Flying Post*, 19 August 1819.
11. *Sun* (London), 28 August 1819.

PART V UNITING THE KINGDOM
1 George IV in Ireland
1. Neil Herman, 'Henry Grattan, the Regency Crisis and the Emergence of a Whig Party in Ireland, 1788–9', in *Irish Historical Studies*, vol 32, no 128 (Nov 2001), p497.
2. *The Globe*, 3 August 1821.
3. National Archives, ADM 51/3252.
4. Duke of Buckingham and Chandos, *Memoirs of the Court of George IV, 1820–1830*, London, 1859, vol 1, p103.
5. *Memoirs of Sir William Knighton*, ed Lady Knighton, vol 1, Paris, 1838, pp94–7.
6. Buckingham, op cit, p107.
7. Knighton, op cit, p98.

2 To Scotland
1. Buckingham, op cit, p193.
2. *The Letters of Sir Walter Scott*, ed Grierson, vol 7, 1934, p213.
3. Ibid, p211.
4. L Cope Cornford, *A Century of Sea Trading*, London, 1924, p72n.
5. *The Sun*, 12 August 1822.
6. Navy Records Society, Boteler, op cit, p101.
7. *Public Ledger and Daily Advertiser*, 17 August 1822.
8. *Memoirs of Sir William Knighton*, op cit, p119.
9. Scott Letters, op cit, p241.

3 The Lord Admiral's Excursion
1. Navy Records Society, Boteler, op cit, pp 97–8.
2. National Archives, ADM 51/3396.
3. *Despatches, Correspondence and Memoranda of Field Marshal Arthur Duke of Wellington*, London, vol 4, 1867–80, p514.
4. Ibid, p535.
5. Ibid, p554.
6. William R O'Byrne, *A Naval Biographical Dictionary*, London, 1849, vol II, p1103; vol I, pp 412–13.
7. *Wellington Correspondence*, vol 4, op cit, pp628, 581–2, 602.
8. National Archives, ADM 87/3.
9. Ibid, ADM 87/3.
10. *Windsor and Eton Express*, 17 May 1834.

PART VI BUILDING AND SAILING
1 Captains and Crews
1. Compiled from Navy Records Society, *Catalogue of the Pepysian Manuscripts*, op cit, vol 1, and John Charnock, *Biographia Navalis*, London, 1797, vols 1 and 2, passim.
2. *Calendar of State Papers, Domestic*, 1696, London, 1913, 16 December 1696.
3. Navy Records Society, *Catalogue of the Pepysian Manuscripts*, op cit, vol 4, p403.
4. Ibid, p224.
5. National Archives, SP 42/15.
6. Ibid, ADM 51/586.
7. Ibid, ADM 106/3301.
8. Navy Records Society, Boteler, op cit, p94.
9. Ibid, pp94–5.
10. National Archives, ADM 106/1104.
11. Ibid, ADM 106/1073/139.
12. Navy Records Society, Boteler, op cit, p226.
13. National Archives, ADM 354/166/122.
14. Ibid, ADM 106/366/248.
15. Ibid, ADM 106/647/97, 98, 651/10, 11.
16. Ibid, ADM 106/740/79.
17. Ibid, PROB 31/92/327.
18. Ibid, ADM 106/777/38.
19. Ibid, ADM 106/360/413. 106/694/191, 354/204/274.
20. Navy Records Society, Boteler, op cit, p93.
21. Goodall diary, op cit, with some discrepancies with 36/12670.
22. National Archives, ADM 39/2097.
23. Goodall, *Small Observations*, op cit.
24. *Daily Courant*, 22 March 1703.
25. Navy Records Society, Boteler, op cit, pp94, 99, 101–2.
26. National Archives, ADM 37/6476.
27. Ibid, ADM 36/12670.

2 Design and Build
1. *British Naval Documents, 1204–1960*, ed Hattendorf et al, 1993, p267.
2. *Deane's Doctrine*, op cit, p52.
3. *Catalogue of the Pepysian Manuscripts*, vol 1, p 295; Deane, p112.
4. National Maritime Museum ship's plans, HIL0147.
5. National Archives, ADM 106/3308.

3 Accommodation
1. See Part IV, Chapter 1.
2. Goodall Diary, op cit.
3. National Archives, ADM 106/3328.
4. Dalton, *British Royal Yachts*, op cit, pp314–15, etc.
5. National Archives, ADM 106/3301.
6. Ibid.
7. National Archives, ADM 106/3309.
8. Ibid, ADM 106/1104.
9. Ibid, ADM 106/3301.
10. *Evening Mail*, 9 August 1822.
11. Navy Records Society, Boteler, op cit, p97.
12. Ibid, p93

4 Art and Decoration
1. Dalton, *British Royal Yachts*, op cit, pp314–15, etc.
2. Navy Records Society, *Catalogue of Pepysian Manuscripts*, vol IV, p520.
3. National Archives, ADM 106/3308.
4. Diary, 18 January 1671.
5. National Archives, ADM 106/3296.
6. Acts of Parliament, *New Churches in London and Westminster*, 9 Anne cap 17, p8.
7. National Archives, ADM 106/3466.
8. Ibid, ADM 106/1055.
9. Dalton, *British Royal Yachts*, op cit, pp314–15, etc.
10. Ibid.
11. National Maritime Museum ship models, SLR0379.
12. National Archives, ADM 106/3317.
13. Ibid, ADM 106/3303.
14. Ibid, ADM 106/3300.
15. Ibid, ADM 106/3303.
16. NMM ZBA3082.
17. Richard Endsor, 'The Van de Velde Paintings for the Royal Yacht *Charlotte*', *Mariner's Mirror*, vol 94, pp271–5; National Maritime Museum oil paintings, BHC3254, BHC 3732, BHC3556.
18. *Catalogue of Pepysian Manuscripts*, op cit, vol 4, p521.
19. National Archives, ADM 106/3317.
20. Ibid, ADM 49/120.
21. Navy Records Society, Boteler, op cit, pp93, 97, 100–1.
22. *Evening Mail*, 9 August 1822.
23. National Maritime Museum ships plans, NDO0768.
24. Ibid, FHD0097.

5 Sailing the Yachts
1. Falconer, *Dictionary of the Marine*, London, 1780, p262.
2. Matthew Tanner, *Royal Yacht Mary*, op cit, p59.
3. Navy Records Society, Samuel Pepys's Naval Minutes, op cit, p220.
4. Dalton, *British Royal Yachts*, op cit, p58.
5. National Archives, ADM 106/3374.
6. Falconer, op cit, p135.
7. *The Willem Van de Velde Drawings in the Boymans-van Beuningen Museum, Rotterdam*, op cit, vol 1, pp63, 66; vol 2, p253.
8. National Archives, ADM 51/685.
9. Dalton, *British Royal Yachts*, op cit, pp81, 92, 103.
10. National Maritime Museum manuscripts, SAN/V/1.
11. National Archives, ADM 1/586.
12. Historical Manuscripts Commission, *Manuscripts of Lady Du Cane*, p27.
13. Navy Records Society, Boteler, op cit, p97.
14. National Archives, ADM 51/586.
15. Ibid, ADM 2/73.
16. Ibid, ADM 1/2106.

PART VII LAST DAYS UNDER SAIL
1 Victoria in Scotland
1. *The Letters of Queen Victoria, 1837–1861*, London, 1907, pp506–7.
2. *Illustrated London News*, 3 September 1842.
3. Royal Collection Trust, *Queen Victoria's Journal*, entry for 29 August 1842, p90.
4. Ibid, p93.
5. *Illustrated London News*, op cit.
6. Journal, op cit, p97.
7. *The Scotsman*, 14 September 1842.
8. Journal, op cit, p98.
9. Ibid, pp98–9.
10. Victoria, Letters, op cit, p588.
11. L Cope Cornford, *A Century of Sea Trading*, London, 1924.
12. *The Scotsman*, 14 September 1842.
13. Journal, op cit, p133.
14. Victoria, Letters, op cit, pp540–1.

Index

Aberlady Bay 113
able seamen 122
Ackworth, Sir Jacob 124
Adam, Charles 120
Adelaide, Queen 117
Admiralty 71, 121:
    Charles II 29, 30, 31, 32, 37, 137
    William III 27, 45, 46, 47, 50, 52,
    Anne 54
    George I 69, 122
    George II 70, 75, 78, 80, 81, 82, 84, 87, 120, 142
    George III 54, 76, 86, 88, 89, 90, 95, 96, 100, 102, 104, 119
    George IV 115, 116
    Victoria 148
Albert, Prince 145, 146, 148
Aldborough 23, 60, 79
Amelia, Princess 87, 96
American revolt 91
Amiens, Treaty of 99
Amsterdam 10, 13, 14, 41, 48,
anchors 140–2
Andrew, Prince 152
Anglesey, Marquess of 106, 108, 110, 120
Anne, Princess 152
Anne, Queen 56
Anson, George Lord 73, 74, 80–7, 120
Aylmer, Matthew 60, 198–9
Ayres, John 34, 36

ballast 15–17, 30, 32–3, 55, 76, 88, 102, 125, 129
Banks, Sir Joseph 88, 90–1, 143–4
Barking 59
Bass Rock 146
Bath, Earl and Countess of 23, 28, 42
Bayeux Tapestry 7
Beach, Sir Richard 18, 30, 31, 34, 43,
Beachy Head 24, 44, 46, 90, 97, 116
Bedford, Duke of 80, 81, 87
Berkeley, Admiral Lord 46, 58, 59
Bermuda rig 17, 152
Berry, Sir John 34–6
Berwick-upon-Tweed 113, 146
Bethune, Marquise de 25
bezaan or bezan yachts 17, 18, 32
Bickerstaff, Sir Charles 32
Bickerton, Captain 88–91
Blackwall 29, 96, 149
Blackwood, Sir Henry 116
Blenheim 57
Bloomfield, Sir Benjamin 109
Blücher, Marshal 104–5

boatswains 121
Boteler, Lieutenant John 100, 112, 120–3, 132, 138, 142
Boult, Captain Richard 30
Boyle, Courtney 122
Boyne, Battle of 44
Bray, Gabriel 122
Brett, Piercy 120, 131
Brielle 25, 38, 46, 48, 49, 56
Brighton 13, 94, 100, 105, 120, 144, 157
Brooking, Charles 133
Browell, William 93, 120
Buccleuch, Duke of 146
Buckler's Hard 89
Burroughs, Thomas 71, 133

Calais 48, 73:
    Allied Sovereigns 105
    Henry VIII 8
    Louis XVIII 104
    passengers to and from 25
    voyages to and from 18, 19, 20, 21, 47, 49, 72, 110, 143
Calshot Castle 89
Canning, George 115–16
Cant Buoy 8
Caroline of Brunswick, Queen 92–3, 107
Canterbury 22, 44, 110
Carmarthen, Earl of 48, 52–4, 110, 124
carpenters 121
Carrickfergus 44, 47
Carteret, Sir Edward 44
Castle, William 27, 124
Catherine of Braganza, Queen 21
Cavendish Spencer, Captain Robert 116, 120
Chapman, Frederic Hendrik af 71
Charleroi 45
Charles I, King 8
Charles II, King 8, 10, 12–14, 17–25, 27–30, 32–4, 38–40, 43–5, 52, 59, 69, 103:
    sea experience 9
Charles, Prince, 'The Young Pretender' 80
Charles, Prince of Wales 153
Charlotte of Mecklenburg-Strelitz, Princess and Queen 83–6
Chatham 31–3, 37, 121:
    boats sent from 22
    Brazier 17
    George IV 110
    importance of 63
    lead at 16, 32, 33
    lighter from 92
    marines 112
    mast from 66
    master shipwright 18
    Peter the Great 48

    Sandwich at 87–92
    ships from 76
    timber from 99, 126
    William III 44
    workmen 133
    yachts built 29, 30, 31, 32, 50, 52, 53, 64
Chichester Harbour 72
Chiswell 95
Christchurch 97
Clements, Captain 25, 119
Cleveland, Duchess of 28
Cleveley, John 133
Clifford, Sir Thomas 38
Cockerell, brazier 17
Codrington, Sir Edward 116
Collier, Sir Francis 145
Collins, Greenville 18, 37, 44, 50
comedians 26
Continental Congress 91
Cook, Captain James 88
Copenhagen 48–9, 64
copper 32, 64, 66, 100–1
Coronelli, Vincenzo 6, 128–31, 134, 136–7, 154
courts martial 18, 20, 36–7, 52, 73, 122
Cowes 99, 106, 108, 115, 117, 149, 151–3
Crane, Dr John 94
Cromwell, Oliver 10
Cromwell, Richard 10
Crow, Thomas 38–9
Cumberland, Duke of 71, 73, 87, 98

Dalkeith 111, 146, 148
Dartmouth 24, 88, 151, 153
Dartmouth, Admiral Lord 42–3
De Ruyter 38–9
Deal 22, 24, 152
Deane, Sir Anthony 14, 16–17, 24–9, 64, 125
Delftshaven 12
Deptford 37, 44, 48, 51–6, 63–73, 82–3, 97–101, 112, 119, 129, 136–7, 144:
    crews at 121, 122, 123
    decorative arts 133, 137
    Duke of Clarence 115–16
    facilities 63–6, 118, 126
    lead 30
    master shipwright 28
    Peter the Great 51–2
    Sandwich at 87–92
    timber from 31
    town 67
    yachts at 20–2, 63–7, 97, 98, 131
    yachts built 27, 53–4, 55, 71, 99, 100, 101, 124, 127, 149
    yacht crews from 75–6

    yachts fitted 33
    yachts repaired 24, 56, 58, 59, 69, 78, 82
Desborough, Captain 55
Dettingen, Battle of 44, 79
Dieppe 20, 21, 25, 26, 72
Dover 24, 43, 62, 72, 90, 106, 116, 143:
    Allied Sovereigns 105
    Henry VIII 7–8
    Louis XVIII 103
    navigation from 68
    passengers to and from 25
    roads 144
    Secret Treaty of 39
    voyages to and from 12, 18, 38, 72
Downs, the 49, 73, 75, 81, 97, 110, 116, 144:
    Duke of Clarence 104
    navigation from 68
    Shovell 4–9
    yachts at 73
dry docks 55, 65–7, 79, 126
Dublin 108–9
Dudman, John 99
Dummer, Edward 50, 65–6
Duncan, Admiral Adam 96
Dungeness 24, 38, 62, 73, 97
Dun Laoghaire (Dunleary, Kingstown) 108–9

East India Dock 112
Eaton, John, seaman 34, 36
Eaton, Thomas, carver 17
Edgcumbe, Admiral Lord 89–90
Edinburgh 77, 114, 146
Edward VIII 152
Effingham, Lord Howard of 115
Elbe 22, 84, 93
Elizabeth of Bohemia 12
Elliot, Midshipman 88
Epping Forest 29
Erith 22, 34, 142
Eugene, Prince of Savoy 67
Evelyn, John 20, 48, 68, 133
Eyles, Captain 104

Falmouth 115
Farne Islands 145
Faseby, Captain William 25, 27, 32
Faversham 43
Field of the Cloth of Gold 7–8
figureheads 8, 50, 92, 133–6
Fitzherbert, Marie 107
flags 8
Flamborough Head 86, 113
Fletcher, Mathias 33
Foley, Admiral 104–5
fore-and-aft rig 139
Forth, Firth of 113
Fox, Uffa 152–3

Frederick, Prince of Wales 77
Fuller, William 48
furniture 130

Garter stars 136
General Steam Navigation Company 147–8
George of Denmark 43, 47
George I, King 54, 58–62, 68, 69, 103, 129
George II, King 44, 69, 77, 80–2
George III, King 83–4, 94–8, 106
George IV, King 92–3, 101, 102, 106, 107–10, 111–14
George V, King 151, 152
George VI, King 152
Gibbons, Grinling 133
gilding 15, 28, 29, 31–3, 71, 83, 90, 133, 136
Gillingham 87
Glorious Revolution 42–3
Gloucester House 94, 97
Goddard, Joseph 123
Goddard, Thomas 121
Gomme, Sir Bernard de 23, 25
Goodall, Joseph 92–3, 123, 141
Goodwin, John 76
Gore, Lieutenant the Honourable Edward 116
Goree 22, 45, 48, 57–8, 61, 78, 79, 81
Goring, Lady 25
Grafton, Duke of 33
Granton 146–8
Gravesend 18, 64, 84, 96–7, 119, 142:
    Charles II 24, 30, 40
    George I 59, 60
    George II 77–82
    George IV 112
    governor of 22
    impressment at 74
    passengers at 68, 72
    Queen Caroline 93
    Queen Mary 43
    racing to and from 20
    Tiger off 32
    William III 44–6
graving 66
Green Cloth, Board of 58, 84,
Greenwich 20, 47–51, 55–65, 67–70, 84, 92–3, 111–15, 123, 137, 145:
    George I 59, 60
    George II 77–9
    George III 96
    George IV 111, 112, 114
    Henry VIII 63
    impressment 73, 74
    moorings at 18, 64, 67
    pensioners 75
    Peter the Great 48

Queen Caroline 93
Queen's House 68
Sandwich at 87, 90
voyages to and from 25, 43, 56, 57, 72, 140, 142
yachts built at 30, 32, 33, 124
Greville, Fulke 95
Gunfleet 38, 42, 46, 75, 77, 81, 142, 144
Gunman, Captain Christopher 22, 34–7, 122
gunners 121

Hague, The (Den Haag) 12, 20, 45, 46, 56, 57
Hamoaze 23, 25, 116
Hanover 61, 62, 71, 78, 80, 82, 121
Hanover, House of 58, 60, 68, 86, 88, 138
Harcourt, Earl of 84, 86
Hardy, Sir Charles 77–80
Harwich 26, 45, 61, 64, 78, 82, 120, 140:
  George II 77
  navigation to 68
  voyages to and from 46, 49, 57, 72, 73, 80
  yachts at 22, 40, 42, 56, 79, 81, 84, 86
Heligoland 84
Hellevoetsluis 21, 52, 57–61, 72–4, 77–82, 105, 140–2
Henderson, D and W 150
Henrietta, Duchess of Orleans 18, 19
Henry VIII, King 7
Henslow, Sir John 99
Herreshoff, Nathaniel 150
Het Loo 45
Hinder 60, 62, 78
Hollar, Wenceslaus 8
Hollesley Bay 81
Holmes, Sir John 22
Holmes, Sir Robert 24
Holyhead 18, 108–9
Holyrood House 111, 113, 148
Hood, Captain Alexander 64
Hopetoun House 111, 114
Howard, Sir Robert 15
Howth 108
Hunloke, Major Philip 151
Hunt, Edward 64
Hurstmonceaux Castle 14
Hyde, Lady Henrietta 25

impressment 73, 75, 82
Inchkeith 113, 146

'J'-class yachts 152
Jacobites 44, 50, 58, 59, 60, 77, 80, 111, 113, 120
James II, King, formerly Duke of York:
  coat of arms 137
  daughters 22
  deposed 43–4, 60
  and Gloucester 34–6
  King 41–2, 46
  Lord High Admiral 115
  marriage 19, 20, 21
  yachts 10, 16
Jenkins, Sir Leoline 26

Jennings, Sir John 60
Jennings, Sir William 27, 43
Jordan, Admiral Sir Joseph 38

ketch rig 140
kilt 112, 114
Knighton, William 110, 113
Kronenburg Castle 25

Lambeth 18, 28, 44
Landguard Fort 23
Le Havre 25
Leake, Captain 34
Ledman, John 17
Legge, George 20, 25
Leith 38, 112, 113, 114, 146, 147, 152, 153
Lemon and Ore banks 34, 37
Lenham, Thomas 68
Limeburner, Captain 75, 121
Lipton, Thomas 150
Liverpool, Lord 115
Lizard 46, 115
London 14, 21, 22, 34, 38, 40, 43, 44, 45, 52, 58, 59–64, 70–3, 77–8, 84, 86, 87, 89, 92, 98, 104, 110, 112, 113, 116, 125:
  carvers at 71
  culture 133
  Great Fire 7
  Leith, service to 147–8
  Lord Mayor of 59
  need to be close to 60, 82, 105, 109
  painters from 100
  panorama at 96
  plague 27
  Queen at 78
  St James's Palace 82
London Bridge 13, 17, 18, 28
London, Tower of 39
Londonderry, Lord 108
Longships rocks 115
Lord High Admiral 115
Lords Justices 45, 46, 60
lords' room 132
Louis XIV, King of France 19, 42, 57
Louis XVIII 103, 116
Luddites 105
Lulworth 95
Luttrell, Narcissus 46
Lyme Bay 24
Lymington 41, 97, 98
Lyndhurst 97

Maastricht 56
MacBride, Admiral 95
Mainwaring, Lieutenant Jemmet 122
Malmesbury, Earl of 93
Malplaquet 57
Margaret, Princess 152
Marlborough, Duke of, John Churchill 22, 36, 46, 56, 57, 69, 73
Mary, Princess of Orange, Queen 21, 43
Mary of Modena, Queen 18–21
masters 121
Medway 15, 27, 38, 40, 53, 55, 66, 79, 88, 90, 128
Melville, Lord 111, 115, 117

midship section 125
Milford Haven 109, 115
Milton and Canot, print 63, 64
Mitchell, Admiral Sir David 46
Molloy, Captain Charles 64, 66, 70, 71, 78–80, 118–20, 143–4
Monmouth, Duke of 18, 23, 41
Montrose, First Duke of 60
Montrose, Third Duke of 111
Morland, Sir Samuel 31
Morrice, G F 116
Morrice, Fig 95
Morrice, Salmon 52
Mount Batten 23, 25
Murry, John 92

Naish, Catherine 83
Napoleon Bonaparte 103, 105, 107, 116, 135
Naval Chronicle 100
Navy Board 27, 30, 31, 38, 46, 55, 64, 75, 76, 89, 104, 120, 121:
  Charles II 15, 17, 28, 29, 32, 133
  William III 50, 51, 52, 53, 65, 119
  Anne 54
  George I 69
  George II 70, 73
  George III 83, 88, 96, 99, 100
Neal, Sir Harry Burrard 96, 97
Needles, The 41, 42, 90, 91, 97, 98, 106, 108, 115, 149, 152
Nelson, Horatio 95, 96, 104, 105, 109, 115, 116, 120
Netherlands (United Provinces) 8
Newark Sand 34
Newcastle, Duke of 80, 154, 156
Newhaven 116
Nore 18, 52, 60, 73, 74, 78, 79, 81, 82, 87, 90, 92, 93, 100, 112, 128, 140, 142, 145:
  anchoring at 38, 40, 48, 59, 75, 80, 97, 113
  buoy of 26, 45
  mutiny 96
  paying ships at 55, 56
  Resolution at 88
Norman, Samuel 83, 85
Normandy 7
North Foreland 22, 24, 48, 61, 62, 80, 97, 116, 143, 144
Nugent, Thomas 60–1

Oaze Edge Buoy 24
O'Connell, Daniel 109
Omai 90
Orange Polder 46, 49, 56, 59
Ordnance, Board of 16, 17, 18, 23, 29, 104
Orfordness 60, 77, 79, 81
Ormond, Marquess of 10
Orrey, sand 34

Osborne House 151
Oudenarde 57
Owers Bank 97

Padworth, Berkshire 7
Paget, Sir Charles 107, 109, 113, 120, 145
paintings 137–8
Parliament, English 9, 26, 27, 76:
  Charles II accepted 12
  King to be close during session 45–6, 60, 78, 96
  representatives for Queenborough 92
  secured for George I 58
  Third Dutch War 41
  Thirty Ships (1677) 25, 30
  William and Mary proclaimed 43
Peake, Sir Henry 100, 102, 124
Peel, Sir Robert 111, 146, 148
Pendennis Castle 115
Penzance 108
Pepys, Samuel 28, 29:
  asks Navy Board for money 27
  Deptford 64
  Edward Dummer 50
  patron of Deane 28, 124
  qualities of ships 53
  relations with King 20, 24, 25, 33, 140
  sailing in yachts 18, 23, 34–7
  ship decoration 133
  on yachts 14, 15, 17, 18, 19, 30, 125
Peter the Great, Tsar of Russia 46, 48, 52
Pett family 124
Pett, Christopher 16, 17, 18, 64
Pett, Peter 14, 15
Pett, Phineas 29–33, 125, 137
petty officers 122–3
Philip, Prince, Duke of Edinburgh 152–3
pilots 142
Placentia 8, 63, 68
Plymouth 23, 25, 63, 87, 88, 89, 90, 91, 115, 116, 117
Poniatowski, Prince 88
Portland 24, 94, 95, 98, 108
Portland, Earl of 52
Portsmouth 6, 23, 34, 44, 66, 95, 104, 106, 109, 119, 121:
  Duke of Clarence 115, 116
  goods carried to 26
  rail connection 149
  Royal Naval College at 100
  royal visits 21, 22, 24, 25, 107
  Sandwich at 87–91, 123
  School of Naval Architecture 124
  soldiers and seamen carried to 27
  stores from 97
  strategic importance 63
  voyages to and from 42, 43, 76
  workmen from 133

yachts at 20, 101
yachts built at 28, 29, 64
Portsmouth, Duchess of 33
pressing see impressment
Prussia, King of 72, 77, 82, 104, 105

Queen's House, Greenwich 8, 68

racing, yacht 16, 18, 20, 92, 144–5, 149, 150, 151, 152, 153
Radical War 111
Ramillies 57
Ramsden, Isabella 84
Ramsgate 24, 110, 112
Redbridge 99
rebuilding 66
repairs, types of 65–6
Richard I, King 7
Richmond, Duke of 72
Ripley, Captain William 52
rolling grounds 46, 51, 84
Rooke, Admiral Sir George 44, 45, 48, 49, 56
Rotherhithe 18, 27, 75, 108, 123, 124, 133
Rotterdam 10
Royal Naval College 87, 100
Royal Naval Hospital 68
Royal Yacht Squadron 149
Ruffhead, John 30
Rupert, Prince 22, 40
Rye 25, 38, 40, 61, 62
Ryswick, Treaty of 47

St Abb's Head 148
St Helens 89, 90, 104, 106
St Michel, Balthasar 24, 29
St Nicholas Island 23, 25
St Vincent, Earl of 96, 112
Sanderson, William 34, 50, 51, 53, 119, 123, 131
Sandwich, First Earl of 21, 22, 25,
Sandwich, Fourth Earl of 70, 71, 80, 81, 87–92, 116, 122, 123, 144
Saumarez, Admiral James 95
Saunderson, William 118, 119
Scarborough 21, 113, 145
Scheveningen 12
Schoonveld 41
Scott, Sir Walter 112–14, 120, 148
Seaforth, Earl of 90
Seppings, Sir Robert 102, 108
servants, status of 123
ships and other vessels, apart from royal yachts:
  America (1851) 149
  Barfleur (1768) 89–90
  Black Eagle (1831) 145
  Centaur (1759) 88
  Charles Galley (1676) 45, 46
  Comet (1812) 108
  Comet (1822) 112, 113, 114, 115
  Fanfan (1666) 28
  Gloucester (1654) 34–7, 122, 144
  Greyhound (1672) 24, 25, 28, 45

INDEX 159

Happy Return (1654) 34, 36
Harwich (1674) 25
Impregnable (1810) 104–6
James Watt (1821) 112–14
Jupiter (1778) 92–3
Liffey (1813) 108–9
Lightning 108, 109, 115, 145
Magnificent (1766) 95
Monkey (bought 1837) 145
Montague (1654) 22
Peregrine Galley (1700) 51, 55, 70, 121: bow 125; design and build 53, 124; figurehead 134; George I 57, 59; keel 126; ownership 54; repair 58; rig 139; sails 142; seamen 123; voyages of 56, 58; William III 119
Pique (1834) 117
Resolution (1654) 21, 22
Resolution (1667) 23
Resolution (1770) 88–90
Royal James (1675) 24, 25
Royal Sovereign (formerly *Sovereign of the Seas*, renamed 1660) 41
Royal Sovereign (1701) 56
Rupert (1666) 23, 38
San Fiorenzo (captured 1794) 96
Southampton (1757) 95
Sovereign of the Seas (1637) 8
Surprise (later *Royal Escape*) 9, 13
Swiftsure (rebuilt 1696) 47–8
Tiger (1681) 18, 32, 142
Trinity Royal (1416) 8
Victory (1765) 70
Warspite (1758) 76
sheer hulk 66
Shish, Jonas 18, 28
Shish, Thomas 32, 124
Shoreham 9
Shovell, Sir Cloudesley 44, 46, 47, 48, 60
Sidmouth, Lord 109
Skinner, John McGregor 109
Slade, Sir Thomas 70
Slough 145
South Foreland 90
Southampton 27, 29, 89, 95, 106, 116, 151
Southwark 59, 148
Southwold Bay 46, 57, 59
Spencer, Earl 96, 105
Spithead Mutiny 96
Stacey, Isaac 121
Stade 60, 84, 85
staircases 128
Stanhope, Earl 60
steering wheel 141
Steinberg, Baron 79
stern decoration 134–5
Sturgeon, William 34, 36, 122
surgeons 121
Swain, Richard 17
Swin Channel 74, 79, 142
Symonds, Sir William 117, 124, 149

tacking 141
tallowing 66

tarpaulin captains 118
Tattersall, Nicholas 9
Test Act 34
Thames Estuary 42, 43, 91, 99, 151:
 chart 37
 guarding 96
 passage 77, 78, 80
 possible naval battle 9
 Red Sand buoy 24
 yachts in 40, 59
Thames, River 17, 63, 69, 86:
 carvers 71
 depicted 65
 Dutch Wars 38
 George I 59
 impressment 75
 mooring 13, 81, 84, 143
 passage 68, 90, 91, 140, 142
 police 114
 racing 18, 20
 river craft 27
 shipyards 99
 steamships 112–13, 147
 yachts built on 64
 yachts in 14, 28, 34, 36, 55, 57, 58, 59
Yacht Club 151
Tilbury 36
tiller 141
timber 125:
 decayed 70
 Deptford wet dock 65
 yachts 14, 16, 52, 102, 117, 126
 supply 17, 29, 30, 32, 66, 68, 69, 89, 99
Townshend, Lord 60, 62
Trinity House 18, 64, 67
Turner, Henry 133, 136, 137

Utrecht, Treaty of 57

Van Ghent 39
Van de Veldes 10, 14, 27, 134, 147:
 decorations for yachts 33, 133, 137
 the Elder 13, 16, 21, 32, 40, 41, 141, 142
 the Younger 11, 13, 15
Versailles 24
Victoria, Queen 117, 145–51
Vile and Cobb 83
Virginia Water 117
Vlooswick, Mr de 1, 3–4

Wager, Sir Charles 78–9
Walford, Captain John 41
Walker, Isaac 17, 18, 31, 32, 33
Wallis, John 125
Walpole, Galfridus 64, 119
Walpole, Horatio 72, 83
Walpole, Sir Robert 77, 78
Warren, Sir John Borlase 95
Warson, G L 149–52
Waterloo 105, 108, 110
weather forecasting 45
Wellington, Duke of 103, 105, 116, 117
Wells, John 99
Werden, John 24, 25
wet docks 63, 66, 67, 71, 97, 119, 126

Weymouth 94–102, 108
Wight, Isle of 24, 25, 89, 108, 144, 152:
 anchored off 99
 Britannia at 151
 Charles I 9
 fleet off 104
 governor 29
 regatta 105
 yachts to 28, 106, 149
Wilhelm II, Kaiser 150
William I, King 7
William III, of Orange, King 20–2, 31, 42–9, 52, 53, 56, 60, 68
William IV, King, formerly Duke of Clarence 95, 103–5, 115–17, 120
Wilshaw, Thomas 33
Windsor 37, 94, 117, 145, 148,
Worcester 9
Worcester, Bishop of 98
Worcester, Marquess of 10
Worthing 96
Wren, Sir Christopher 68
Wyborne, Captain 34, 36
Wynn, Charles 109, 111

yachts:
 characteristics and types 8–9, 33
 construction 126–7
 design 124–5
 hull lines 125
 launch 127
 origin of 8
yachts, named:
 Anne (1661) 121: building 16–17, 64, 123, 124; captains 25, 36, 118, 119; decorations 137; dimensions 126; officers 121; racing 18, 20; voyages 21, 22, 23, 24, 36, 38, 40, 64, 125, 137
 Augusta (ex *Princess Charlotte*, 1771): altered 69, 71; captains 120; crew 123; officers 122; plans 130; renamed 55, 71; rig 140; sailing 141–4; structure and layout 126, 128; voyages 84, 85, 88–93, 96–101, 107
 Bezan (1661) 17, 18, 26, 139
 Bloodhound (1936, owned from 1962) 153
 Bluebottle (1948) 152–3
 Britannia (1893) 149–52
 Britannia (1953) 6, 153
 Carolina (ex *Peregrine Galley*, renamed 1716) 51, 54, 62, 65, 67: accident 64; cabins 133; crew 71; officers 121; renamed 59; stern 129; voyages 61; use by King 60, 77
 Charles (1662) 18, 25
 Charles (1675) 20, 29, 32, 64
 Charlotte (1677) 30–2, 34, 36, 37, 44, 46, 49, 59, 64, 67, 73, 74, 75, 77–83
 Cleveland (1671) 15, 20,

23, 25–9, 39–43, 67, 69, 118–21, 139, 142: captains 25, 118, 119, 121; decoration 32; Deptford 64; design and build 28, 124; repair 55, 57; stern 134; use by King 29, 40, 41; voyages 26, 27, 42, 43, 45
 Coweslip (1949) 153
 Elfin (1848) 149
 Fairy (1844) 149
 Fubbs (1682, rebuilt, 1724) 47, 55–7, 66–7, 69–70, 72–6, 78–81, 89–90, 133–4: boat 133; cabins 130, 131; crew 76, 78; design and build 32, 124, 125; goods carried 26; head 134; impressment 73–5; naming of 27; officers 121, 122; repair and rebuild 55–6, 66, 69, 70; rig 139, 140, 141; sailing 33, 62, 143–4; scrapped 90; use by King 50; voyages 43, 45, 46, 48, 57, 60, 72–4, 79, 80, 83–5, 89
 Henrietta (1663) 17, 22, 23, 29, 36, 41, 119, 123, 124
 Henrietta (1679) 26, 56: anchors 140; captains 119; design and build 32; repair 55; sailing 142; stern 125; voyages 44–8, 58, 59, 60
 Isabella 31, 32–3, 46, 69
 Isabella Bezan (1680) 32, 33, 139
 Isle of Wight 29
 Jamie or *Jemmy* (1662) 18, 124
 Katherine (1661) 14–18, 39, 134: captains 25, 119, 121; cost 133; design and build 14, 64, 124; interior 128; racing 18, 20; voyages 22, 38, 42, 43, 45, 48, 49, 59, 60
 Katherine (1674) 26, 42, 43, 45, 48, 49, 59, 60, 67: anchors 140; crew 76; design and build 30–1, 125; impressment 38, 75; repair 69, 90; replacement 100; voyages 22, 24, 34, 36, 72, 73, 78–85
 Kitchen (1670) 25–8, 41, 42, 64, 119, 124, 133, 134, 142
 Mary (1660) 13, 14, 16, 17, 18, 29, 125
 Mary (1677) 26, 64, 67: captains 37; design and build 31; repair 55; rig 33, 69; voyages 20, 21, 22, 34, 36, 43–8, 49, 55–60, 72
 Merlin (1666) 18, 21, 27, 38, 39, 41, 64, 133, 134
 Monmouth (1666) 24, 29, 38, 119, 133
 Navy (1673) 20, 29

Portsmouth (1674) 20, 24, 26, 44, 124, 137
Prince Regent (1820) 100, 113, 120, 124, 126, 132, 141
Princess Charlotte (1711) 55
Queenborough (1671) 29, 75
Royal Adelaide (1833) 117, 124, 127, 149, 153
Royal Caroline, or *Royal Carolina* (ex *Peregrine Galley*, renamed 1733) 66, 71, 79–81, 136, 137, 142
Royal Caroline, or *Royal Carolina* (1749): cabins 131; decoration 133, 136, 137; design and build 70–1; drafts 127, 129; crew 71, 74, 76; impressment 73, 75; officers 121; replacement 87; voyages 83, 128
Royal Charlotte (renamed 1761) 86, 90, 96–100, 102, 104, 105, 123
Royal Charlotte (1824) 132, 138, 149
Royal Escape (adopted 1660) 10, 13
Royal George (1817) 100, 126, 135: cabins 131, 132; captains 120, 121; compared with steam yachts 148–9; crew 123; decoration 138; depicted 99, 101, 111; design and build 124, 125; figurehead 15; interior 128; launch 105; officers 122; rig 142; steering 141; use by King 102, 105, 109, 110–19; Queen Victoria 145–7
Royal Sovereign (1804) 108, 129: anchors 140; bow 125; bulkheads 136; cabins 131–2; captains 120; coppering 66, crew 121, 123; decoration 138; design and build 98–104; interior 128, 130; launch 127; officers 122; sailing 142–3; steering 141; voyages 103, 105, 109, 112, 113, 116
Royal Transport (1695) 46, 52, 53
Saudadoes (1670) 19, 22, 24, 25, 28, 42, 45
Victoria and Albert: I (1843) 49; *II* (1855) 149, 151; *III* (1899) 151
William and Mary (1694) 46, 50–1, 53, 57, 64, 66, 67, 69, 72, 73, 75, 77–82, 87, 89, 90, 100
Yarmouth 24, 34, 46, 64, 90, 114